AMBERGRIS

By

Douglas Haydon

This book is a work of fiction. Places, events, and situations in this story are purely fictional. Any resemblance to actual persons, living or dead, is coincidental.

ISBN: 1-4140-2470-3 (e-book)
ISBN: 1-4140-2469-X (Paperback)
ISBN: 1-4140-2468-1 (Dust Jacket)

Library of Congress Control Number: 2003098167
.

This book is printed on acid free paper.

Printed in the United States of America
Bloomington, IN

1stBooks - rev. 11/12/03

For Sharon, the love of my life, my wife and soul mate, who has guided me with ample love, support, and good sense.

A special thanks to my parents, Randall and Priscilla, for an upbringing fertile to a restless imagination.

Thanks to Steve Holderman for the fantastic artwork that adorns this book. Steve can be reached at:

sholderman@sbcglobal.net

The author can be reached at:

AmbergrisNovel@aol.com

Prologue

He turned his back to the morning sun, rotated the 28-105 mm zoom lens on the Nikon until the outstretched beach in the crosshairs sharpened into focus, and squinted to read the illuminated icons at the base of the viewfinder. Everything was ready to go. He placed the camera in his young son's hands, instructing the boy to secure the strap around his neck and bring the viewfinder to his eye. The boy followed his father's finger to a spot on the horizon. After a few minutes of instruction on aperture, he gave the order, and the boy clicked-off a picture. He returned it to his father, receiving a pat on the back before the two began walking up the beach.

The boy traced a serpentine pattern of seaweed carried in by an early morning high tide. Cold winds from the east flapped beachside eelgrass, lifting fine salt-and-pepper sand, crusted by the sun, airborne; it continued up shore, tossing isles of warning-red poison ivy, and dashing into the wooded heart of the peninsula. He pulled the camera to his eye, watching his son hopscotch over a dike of barnacle-clad boulders. He snapped off a few, and instructed his son not to journey too far out on the dike.

The boy crawled to the top of a craggy rock and descended its hidden face, disappearing to the other side of the dike. The father shouted to the boy demanding he wait for him. With care, he looked for and found an easier passage than the boy had taken over the dike.

"Look, Dad," the boy said as his father leapt to the sand beside him.

The father squinted in the direction his son was pointing down the beach. The father pulled the Nikon to his eye and zoomed-in on the general area. First glimpse brought a white disheveled mass in rapid movement floating above the beach. He zoomed in closer. "They're seagulls." He took a cloth from his front pocket and polished the lens.

"There must be millions of them."

"Let's take a closer look." He squatted and instructed the boy to mount his shoulders. "You're our scout."

The boy scrambled to his wide shoulders and the two moved in the direction of the disturbance. As they drew closer, the cackle from the horde amplified.

"Eee-ou." The boy jerked his sand-covered hand over his nose and mouth. "It stinks," he said his voice muffled.

The father took a few deep breaths and detected a fetid strand woven in the briny air. Something was dead. And the something was feeding the gluttonous bevy of feathers in the air above. He lowered his son from his shoulders. "Wait here, son."

"Why?"

"Because," he said firmly.

"I never get to go..."

"Let me take a closer look. If everything is okay, you can join me."

The boy fell to the sand in a thud, grabbed a handful, flung it down with force, and pouted. The father felt his blood pressure rising. He wasn't sure if he wanted to pursue this further. He recalled in his youth, the trip to Canada, the accident, it still lingered in his mind, Dad driving by and telling him not to look. He looked. And he didn't want to look again. He shook the image free and raised the camera to his eye. An oblong dark gray object with white spots lay center beneath the gulls, some having lighted on it. He released the camera and rested it against his belly.

As he drew within a few hundred feet the seagulls perched atop the mass flew up to be absorbed in the chaos. Waves of nausea, stirred by the tremendous stench, began a retching sequence in his belly. With his right hand he pulled the lens cloth from his front pocket and placed it over his nose. With his left hand he brought the camera up for a closer look. He moved the crosshairs along the curvaceous object—back and forth—twice. He studied one end carefully and concluded he was looking at a tail. Damn, that's a big fish. Must be a whale. Turning around and looking at the boy, he found him darting up and down the beach and sliding into the sand, but maintaining his distance.

The odor remained constant, thwarting even the wind, continually fed by the beast, exuding gas, squeezed out by its own tremendous mass. The lens cloth was pulled snug to his nose as he marched forward. The gulls complained in force, an intimidating chorus of sonorous cries filling the skies at alarming levels.

A Humpback, he concluded when standing within a few feet. The arid wrinkled corpse, gritty with stubbles of sand, and caked in bird feces, bore no similarities to the creatures he had observed during a whale-sighting trip out of Hyannis last summer. The creatures enthralled him and the group at starboard as they breached the ocean surface, soaring in the air, and falling back to the sea in a thunderous clap. He pitied the whale for having succumbed to what he considered an unbefitting death for such a majestic creature.

The oppressive teaming of smell and sound began strangling his senses. He decided to take a quick look around the creature and make a hasty departure back to his son. Starting at the tail, he observed something he missed at a distance. Exposed, like the vertebrae of a dinosaur in the badlands of South Dakota, the tight coils of a thick rope, secured at one end around the flukes of the whale, and at the other, a knotty oak tree at the periphery of the woods. It occurred to him that Park Rangers, or some other group acting in an official capacity, had probably been by and taken action. He determined that the authorities had some practical reason for not allowing the carcass to return to sea.

As he walked around the tail he studied the thick rope. It had made deep cuts into the blubber. He wondered how this could have happened if the rope had been attached after the whale expired. Perhaps the motion of the ebb and tide of the sea had generated the necessary friction. He walked around to the head. A gelled eye, black and lifeless, reflected the gulls. He bent to his knees and observed the long mouth, its shape forming an eerie, yet natural smile. He leaned forward to look at the strange baleen sprouting from plates in the whale's mouth. The majority was encrusted and steadfast, but a few fibers remained loose to toss in the wind. Intermixed in the dark gray hair-like baleen were finer strands of brown. These were not organized at the roof of the mouth like the gray fibers. A burst of wind howled from the east, infiltrating passages into the whale, and a lone sweet scent drifted from behind the wall of baleen. He felt a strange compulsion to lift the baleen and feel it in his hands. As a tuft rested in his palm, he rolled the individual fibers between his fingers. The brown strands felt very fine to his touch. Driven by ever more curiosity, he lifted the baleen and looked behind—

He fell back and screamed, "Oh my gosh."

The boy had surreptitiously crept close.

"Get out of here, Brett," he yelled as he backpedaled, sand spraying beneath his feet. "Dear Lord..." He fell into the surf.

The boy froze.

From the beach, opposite his approach to the carcass, a man was running toward them.

"We gotta go...we gotta—" The father recovered from the surf—his trousers drenched—grabbed his son, hoisted him to his shoulders, and ran in the direction of the dike.

From behind: "Hoosier!"

Ambergris, am'ber-gres, *n.* A solid, opaque, ash-colored inflammable substance used in perfumery. It is a morbid secretion obtained from the spermaceti whale.

Chapter I

Earlier…

He pressed his solid frame against them, swinging open the glass doors at Pequod Corporation, and stepped out on the sidewalk at William and 8th. The sky, cloaked in a dynamic patchwork of gray and white frayed clouds, encased the city of New Bedford, Massachusetts in a mizzle. A cold wind off the harbor brought a tingle to Bennett Ackerman's face and teased his fine dark brown hair, feathering it at the sides. He pulled his coat snug around him. On the other side of William a mixed group paraded, led by a silver-capped matriarch, chatting sporadically between puddles, each member gripping some form of an umbrella. The dampened smell of fresh fish-and-chips from Cousin's Seafood traversed from house to building to house and so on, bringing a mild comfort to an unusually chilled early May afternoon. Down the road a crowd gathered to celebrate Maritime Heritage Day. The umbrellas were heading in that direction.

Inside, Bennett ached for summer. The winter had been long and uncharacteristically laden with treacherous winter storms, most of which dropped sheets of ice from rooftops to roads. The ice crippled every component of industry and technology, including telecommunications. As Chief Engineer at Pequod, an Internet Service Provider (ISP), Bennett had witnessed system failures here and there over the years, but not ten in a span of three months. He reached 6th Street and jumped from dry spot to dry spot until landing on the other side. A yelp came from across the street as someone in the pack slipped, but was caught by nearby members. The ten unplanned outages had cost Pequod customers.

The green and white flag above O'Malley's Bookstore hung weighted with moisture, not flexing in the breeze, but swinging as a whole.

When Bennett stepped inside, he was greeted by the tantalizing scent of newly hatched books. He walked up to the cashier, stationed inside a dark wooden circular desk, and pulled a *Computer Shopper* from a pile of the same, surrounded by stacks of other computer magazines. A young freckled boy placed a comic book on the desk and was rummaging through his pocket for money while the cashier read the cover of the comic book. The boy produced a single dingy quarter. The cashier started to explain to the boy that the quarter would not be enough for the comic book, when Bennett grabbed his attention by pointing to himself and mouthing that he'd cover it. The cashier smiled and changed his story midstream, explaining to the boy the comic book was covered.

At Pleasant Street, the towering public library—outlined by a black wrought iron fence and accentuated by towering spires of red brick, some coated in a pallid green moss—implanted its gothic architecture near the bay. Bennett looked to his right at the picturesque library. He recalled his hours spent there in his youth. He and his best friend Wade, poring over musty books on fish, the scientists photographed inside wearing thick black-rimmed glasses and crew cuts, and he and Wade dreaming of the day when they would be ichthyologists with their own ship, sailing the warm seas of the Caribbean, diving and finding all the mysteries the depths. Their interest so enthusiastic, they had written reports and made their own 8mm films, all without the slightest nudge from teachers or parents. Bennett felt a tinge of anguish. He tried to fight it off, to ignore it, but he knew better. When he delved, as he did now, he knew in retrospect, the motivation sprung less from curiosity then from his inexorable quest to please his father and thereby gain his acceptance, something that never could occur, his father's life having been taken by the sea. He allowed a tear—who could tell—his face dripping from moisture. That was a long time ago, enough.

Nimble steps across Purchase Street and a right at Acushnet, heading into a line, on his right, of federal style buildings at the waterfront, impressed him with the ornate lintels and balustrades on their uniforms of red brick. The wind now struck from his left, unabated by buildings, nearly blowing his thin pewter rimmed glasses off his nose. He removed them, placing them in his shirt pocket underneath his coat. He couldn't help but feel a sense of doom. Maybe he could attribute it to the weather and the long arduous havoc it had brought to him, but things had not gone well lately, in many ways, and John talking to the press tonight seemed sealed to the same fate. As president of the New Bedford Historical Whaling Society, or NBHWS as members called it, John proved to be an able leader, generous, knowledgeable, dedicated, imaginative. A dentist by trade, John understood how to deal with people, all sorts, and make them feel more comfortable.

But Kathy Rueger was not like any person John had ever dealt with. Pretty, self-confident, viewers would say arrogant, and willing to sell her soul to make it to the top, to be a national newscaster. She videotaped tough interviews broadcast on the national news or national news shows, and studied them like a professional coach would study tapes of his future opponents—looking for weaknesses in the interviewee and the interviewer. She would tear John apart with the claws extruding from her deft questions.

Bennett reached Union and jumped inside the Uncas Pub. In the darkness he pawed for a hook he knew resided on the wall to his left. After he found the hook and rested his jacket on it, he scanned the room, seeing merely shadows between the neon beer signs, the faint light above the pool table, the cigarette machine, and a couple of video games. Cigarette smoke lathered the air—a permanent fixture at the Uncas. A corpulent figure with a round head, resting on a barstool with a cigar in hand, caught his attention. Bennett walked in the direction and when within a few feet, recognized Collin Knowles. Collin had a half-full draft in front of him and was chatting with a woman sitting to his left. Bennett patted Collin's spongy blonde hair.

"What tha!" Collin spun around on his barstool. He held a half-smoked stogie in his left hand and a half-pound of peanuts in his right. He was a big guy with a big heart. He didn't have the stereotypical appearance of a network operations manager—skinny with a techno-geek flare—but he was such a manager at Pequod. "Bennett, how many times have I told you not to do *that*!" He shot some of the peanuts in his mouth.

Thousands, Bennett thought, but it wouldn't stop him. "So what's up?"

Collin finished chewing. "I'm surprised you made it."

"Look just because I think this woman is gonna beat the crap out of us, it doesn't mean I'm not gonna support our group."

"Oh ye of little faith," Collin said smirking. He turned to the bartender. "Jimmy, turn to channel four."

The bartender wiped his hands on a towel and reached for the set mounted high on the wall by standing on his toes. He clicked the dial to channel four.

"They invented remote control over 20 years ago, Jimmy," Collin kidded.

"When you idiots stop tearing this place apart every Friday night, maybe I can afford a TV with a remote control," the bartender curtly fired back.

"—Its starting. Shhh…" Bennett said, pointing to the set.

A woman's square face and puffy blonde hair filled the 21-inch screen. In light makeup, she had no reason to hide her youth.

3

"…Good evening. I'm Kathy Rueger and this is Kathy's Korner. Tonight my guest is John Phillips, president of the New Bedford Historical Whaling Society." A hungry smile. "Welcome John."

"Thank-you Kathy, it's my pleasure to be here." John smiled, affirming his words.

"Why don't we start with some background on the New Bedford Historical Whaling Society—what's its charter?"

Bennett formed a claw with his hand and waved it in front of Collin's face. Collin brushed his hand aside and flashed Bennett a look of annoyance.

"Well Kathy, the purpose of the New Bedford Historical Whaling Society is to further knowledge of the American whaling industry and New Bedford's regional history."

"What does that translate into?"

"Free admission to the museum—"

"The New Bedford Whaling Museum?"

"Yes. A free subscription to our calendar and newsletter, invitations to exhibits lectures, free use of our research library, invitations to members-only events—"

"Okay John, I think that sums it up pretty well. Let's move on. Why would whaling be something we should continue to honor in this day and age? After all, thousands of whales were slaughtered."

Oh boy, here it comes, thought Bennett.

John smiled guardedly. "Well Kathy, the purpose of our organization is aimed more towards the culture surrounding whaling, not the act of whaling itself."

"Would it be acceptable for me to have an Atlanta Historical Slave Society?"

"What?"

"Well, we wouldn't be focusing on slavery, but the culture surrounding slavery. Is that okay?"

John raised one eyebrow in a questioning slant. The camera zoomed to a close up of his face. Sweat beaded profusely on his brow. The camera pulled back. Kathy's eyes squinted like a predator. John rubbed his hands along the tops of his thighs and said in a shaky voice, "It's a free country—"

"I asked you John. What do you think?"

John re-positioned himself in his chair. "Look, they're not the same thing."

"Don't we all realize now that both were very wrong?"

"Yes, but it wasn't wrong back then. People needed oil and other products that they had no other means of acquiring."

4

"The same could be said for slavery. Where else might a plantation get such cheap labor?"

"Y'all this is bull crap!" Collin yelled.

"Kathy, there is a rich and robust history behind—"

"With slavery as well. We don't celebrate it—"

"We're not celebrating it. We're remembering the significant role it played in the history of our community."

"Very well, John. I can see we're not going to resolve this in the short time we have tonight. Now let me move on to my next concern. In all of our conversation tonight you qualify your organization as historical. Are you really only a historical organization?"

John returned a puzzled look. "What do you mean by that question?"

"Well quite frankly, do some of your members still practice the lost art of whaling?"

John's jaw dropped. "What! *Never*! Some of our members are also members of Greenpeace! We have never discussed, let alone entertained, the concept of actively whaling. You're way off base here, Kathy." John shook his head defiantly.

"Did you know that last week the body of a dead whale was discovered floating in Buzzards Bay, and that an autopsy revealed three deep harpoon wounds in its back?"

Bennett and Collin looked at each other in complete disbelief. "What in the world is she talking about?" Bennett asked rhetorically.

"Miss Rueger," John began in a cautious tone. "I hope you're not insinuating our organization played any kind of a role in this tragedy you speak of?"

"He's too kind," Collin interjected, still shaking his head in disbelief. "I'd told her where to go."

"Well?" she asked, now at the edge of her chair.

"Can we skip the sensationalism and discuss what I came here to discuss—the New Bedford Historical Whaling Society?"

Kathy Rueger turned a cold eye on John. "I'm afraid we're out of time—"

The bartender switched the channel to the Red Sox's training camp game.

"That was not good," Bennett commented.

"Well Bennett, that answers one question," Collin stated matter-of-factly.

"What's that?"

"We know why she wanted us on the show." Collin chugged the remainder of his draft and dropped the mug to the bar like a judge a gavel to a judge's stand. "Until tomorrow my friend."

"Calling it a night already?"

"That interview put a crimp on things, don't you think?"

"Ignore her Collin…everybody else does."

Collin stepped down from his barstool and started his way to the door. He stopped halfway and smiled. "I'm happy as a dead pig in the sunshine."

"What?"

Collin shrugged. "Means I don't care what she thinks." He pulled a cap from the rack and put it on. "So be cheery, my lads," he barked out. "Let your hearts never fail, while the bold harpooner is striking the whale!"

Though he and Collin—two years his senior—had been in close geographic proximity for the latter part of their youth, a friendship had escaped them until they were both working for Pequod, a good twelve years later; the age gap of two years—an equivalent of ten years in post-teen years—was the likely culprit. Bennett found his chubby network operations manager amazing. Collin had memorized many of the early songs once on the tongue of whalers. At any fitting situation he became the great orator, rolling the appropriate lyrics in step with the events at hand. He loved to act, admitting to Bennett one day at Pequod that the stage consumed all his time in college—the computer lab lucky to receive scraps. Collin might well have been a professional actor, Bennett often thought.

Bennett left the Uncas with his senses intact. The bells from marker buoys rang out from the harbor. The wind had calmed and the rain dissipated. The smell of the sea settled in at the wharf. Harpooning a whale? Who would harpoon a whale? Who could harm such magnificent creatures?

Gasoline cascaded out from the opening in the red oblong tank onto the deck of the boat. Paul Kitt thrust his wiry, veiny, tattooed arm out and gave the tank the finger, all the while cussing.

"That's right, blame the gas tank you moron," Patrick Arnold admonished in a controlled voice. He stood propped against the boat's steering wheel. He wore a Harley Davidson cap—his black hair flared in a ponytail out the rear—and three days worth of an aggressive beard. Beneath the stark shadow of the bill, his blood-shot eyes peered down at Paul. Paul took a seat on the edge of the boat. His gaunt square head, stringy blonde hair, and matching sparse mustache looked surreal in the fluorescent lamp humming over his head. The smell of gas permeated the air. Paul started to light up a cigarette.

"Knock it off!" Patrick blasted out.

Paul gave him a what-the-hells-the-matter-with-you look.

"Gas! Cigarette! Helll-ohh!"

Paul shrugged. "I smoke around gas fumes all the time."

"Not on my boat." Patrick motioned Paul towards the dock.

Paul whispered something under his breath, got up, and moved dockside. His smoke billowed into the damp night, adding a strange warmth to the air. Patrick gauged his environment with a cautious eye—quiet; no wind; plenty of stars, but ineffective for lighting. No moon. Typical 2:00 a.m. conditions. Rock music echoed from the bay—a party somewhere, probably miles out. Sound carried across still seas as efficiently as it did copper. Patrick took inventory: a bucket of chopped mackerel, check; fishing poles, check; depth finder, check; underwater microphone, check. Everything would point to an early morning fishing trip. His eyes moved to the dock. At Paul's feet were two long objects. Three-quarters-length of thick rounded wood, one-quarter-length iron rod.

Two harpoons, toggle barbs, Lewis Temple design. Check.

"Get rid of the cigarette. It's time to go," Patrick ordered. Paul took two quick drags and sent the remaining butt sizzling into the water. He grabbed the harpoons from the pier and jumped to the deck of the boat. He lifted a fiberglass door on the transom and carefully slid the harpoons underneath it. They clanged against previously stowed razor-sharp lances with heart-shaped points, stretching to occupy the entire length of their 8-foot domain.

"I'm stoked, man," Paul blurted out after returning the panel to its flush position in the deck. "This is such a rush!"

Patrick grinned. "You're sick, man."

"And you don't get a rush out of this?"

Patrick turned the key and started his dual 300 horsepower Envinrudes. "Money, Paul. It's for the money."

The *Where II*, a 26 foot Boston Whaler, pulled from the dock and headed out to sea. It gradually faded as it moved away from the light of the dock lamp. Seconds later, the engines roared as Patrick gunned it.

Bennett rolled over, half-awake, exhumed from a deep sleep. For a split second he found himself in a panic. The bedside clock read 2:28 am. Something had dragged him from an unconscious state. Three sharp beeps coming from the bedside table cleared the cobwebs from his mind. He fumbled for the switch at the base of the lamp and retrieved his radiophone. The LCD displayed *Adam* and *Alert*. Adam Stevens was the on duty operations officer at Pequod. There must be trouble. Adam wouldn't call unless it was something serious. Bennett held down the button on the side of the phone and got a tone indicating the line was open. He kept the button depressed and spoke.

"Adam, this is Bennett, you copy?" He released the button.

There was a moment of silence, a beep, and Adam's voice.

"Hey Bennett, sorry to bother you. I think we've had an intrusion."

"How serious? What's the damage?"

"Still assessing. So far the only sign of our intruder is a file placed in the root directory on Corillian."

Probably so he can boast he was there. "Can you read the file?"

"Nope, it's binary not text. Erin extracted what text was there, but it's just the standard stuff."

Bennett sighed. Should he go in and investigate or not? He knew it was probably some high school punk trying to prove himself. But then you never really know. Bill, his boss, took these seriously...

He depressed the button. "All right Adam. I'll be in soon. Contact the rest of the emergency response team."

When Bennett reached the operations center at Pequod he could immediately ascertain who worked night shift and who was there as a result of the intrusion. The bright fluorescent lights in the almost Arctic white lab illuminated everyone's state of mind. Most in the lab looked somewhat conscious except for the two men standing by the lab door. Collin's hair was flat on one side giving his head the shape of a half-circle. His eyes were puffy and strained, his clothes disheveled. Next to Collin stood his Deputy of Operations, Will Markoff. Will looked the better of the two; his hair was combed and his shirt tucked-in, but he too had puffy eyes. Will almost always maintained a pristine state of appearance, so his current look of disarray, though most might be proud to have it, by his standards was grungy. Both were slurping coffee from their Cisco mugs.

Bennett moved over to Collin and Will. They chatted a little and then he motioned Adam to join them.

"So let's hear the report, Adam, my man," Bennett requested, trying to pull himself from the fog that clouded his mind minutes ago when he slept.

Adam pulled the clipboard from his side and held it up to read. He scanned the first page and then flipped it back behind the clipboard. He scanned the page behind it.

"Okay, here's where the activity starts," he began. "At approximately 2:00 am, *Wolf Spider*—"

"*Wolf Spider*?" Bennett questioned.

"Yeah, *Wolf Spider*, the program Simon wrote," Adam said.

8

Simon, a black college aged kid, was a brilliant programmer. He was the sharpest coder on staff. Only Bennett could claim to have equal skills. His studies included the source code for most operating systems, at least those he could get his hands on. He had an insatiable desire to understand how everything worked and if that meant going to the original building blocks, then he delved into them. Bennett liked Simon; he was sharp like himself, but not arrogant about it, and in fact downright humble. "Tell me more," Bennett said.

"It's really cool," Adam said with a smirk on his face. "You know how a wolf spider operates. It conceals itself in a hole in the ground. When prey gets close, the spider springs from its hole, grabs the prey, paralyzes it, and takes it down into the hole to eat for supper." Adam demonstrated this with his hands. "Simon's *Wolf Spider* is very similar. It hides itself within the terrain of our network. As it hides it memorizes the terrain. If anything alien begins to meander into our terrain, either overtly or covertly, *Wolf Spider* springs from its hiding and grabs it. It wraps it up to *paralyze* it and takes it to its hole." Adam pointed to a Sun machine in the rack. "Once the prey is in the hole, the hole is disconnected from the network so the prey can't escape or cry for help. If the prey is small enough, it gets moved to a floppy, and then the floppy is ejected." Adam held up a floppy. "Our little buddy is trapped in here."

Bennett shook his head in amazement. "You'd almost think we were talking about sentient beings here, not binary bits."

"Adam gets a little carried away in his descriptions—trust me y'all, we're talking nothing but binary bits," Collin said in groggy voice.

"It's very fast, watch this," Adam said, his voice not skipping an enthusiastic beat. He moved back to a Sun station and inserted a floppy disk. "I'm going to copy a file from this floppy onto this machine." Adam typed in the command to copy a file over. In less than a second a floppy ejected from the Sun station known as the hole. "The file is now trapped. Not bad, huh?" Adam made a Stoogesque gesture common to the ops team, snapping his fingers on both hands and then pounding one into the other.

Bennett held a thoughtful look and then asked, "Slick. Will it do the same with processes?"

"No sweat," Adam said.

"So has *Wolf Spider* delivered anything else to the hole?" Will asked.

"Nope," Adam said.

"Shall we take a look at this puppy?" Bennett asked as he retrieved the floppy from Adam.

"Where's an isolated development machine, Adam?" Collin asked. "This thing may blow a machine and I don't want a ripple-effect on our network."

Adam pointed to a decrepit machine off in a corner of the lab. The keyboard looked like the teeth of a hockey player, half were missing. The monitor displayed psychedelic waves of color behind the text.

"Well, maybe what we've got here will put this thing out of its misery," Collin said, and everyone chuckled.

"Well here goes nothing," Bennett said, and slid the floppy into the drive. Adam jokingly covered his ears. Bennett listed the files on the floppy. Nothing showed up.

"List it so that hidden files are displayed," Will said.

Bennett issued the appropriate command.

```
-rwx-------   1 backerma syseng     72124 May 07 02:02   .whale
```

"Congratulations Bennett, you own the file!" Collin laughed out.

Bennett's login was *backerma*.

"Do you sleep walk?" Adam slipped in.

"Very funny," Bennett said, and gave Adam a shove. "Obviously someone's screwing around with me."

"Or," Will spoke-up, "they just randomly picked you."

Collin pulled a cigar from his shirt pocket and began rotating it in his mouth. He knew he couldn't light up in the lab, but he still got some satisfaction from it. "What's with the funky name—dot-whale?"

"It's appropriate for us, with our name and all," Will said.

"I'm wondering if it has anything to do with the interview with John last night?" Bennett asked.

Collin wrinkled his brow and said, "Well, where's the connection?"

"Think about it, Collin, old boy," Bennett said. "Almost all of us belong to the organization. Whoever sent us this file was sophisticated enough to slip through our firewall, so isn't it conceivable they were intelligent enough to make the connection?"

Everyone pondered the thought. Collin put the cigar back in his mouth and then removed it. "Well, who knows? Look, we don't even know what this file is—"

"Let's run it," Adam said.

"We don't know what it will do when executed," Will said.

"What can it hurt?" Adam replied. He moved to the keyboard and typed in the file name at the command line. "Hit enter?" he asked.

Bennett shot a look at Collin who shrugged and said, "Well, hit enter—"

"Just a second," Bennett interrupted. "Re-enter it so that there's a trace on that file."

Collin gave Bennett a pleased look. "Good thinking, lad."

Adam re-entered the command to include a trace and tapped the enter key and everyone else looked on. The dot-whale file appeared to be running, but the command prompt returned with nothing happening.

"Reckon it's playing possum with us?" Collin asked.

"What does the trace say?" Bennett asked.

Adam brought up the trace file in a viewer. Everyone crowded around to read it.

Bennett spoke-up first. "It tried to open a connection on the network, that failed, and then it quit."

"In other words, phoning home," Adam said.

"More than likely," Bennett said.

"Should we let it?" Adam asked.

Collin shook his head first, followed by the others. He said, "Not until we have a controlled environment with a sniffer on the line to give us all the juicy details about who it's trying to communicate with and exactly what it's trying to say."

Collin ejected the floppy from the Sun station and handed it to Bennett. "Hold on to this. We've got the fox out of the hen house. I suggest we adjourn until tomorrow. All in favor say 'Aye'."

Special Agent Mark DeWald finished tying the laces on his running shoes and sprinted from the hardtop parking lot to the cushy white sand on the beach, his flat black hair bouncing in rhythm with each stride of his toned legs. He had made a New Year's resolution to jog every morning and somehow it stuck. He grimaced as the cold, early morning air from up north stung his face and made his teeth feel hollow. He passed off the annoyance by assuring himself that in awhile his body would generate enough heat to take away the sting. Despite the chill he enjoyed running this time of the season; no distractions, no beach blankets, volleyballs, dogs, pretty women in bikinis—just solitude, an environment unobtrusive to clear thinking. He didn't need distractions; he got enough of those from work. But he needed to think about work, to think about the case.

The edge from the cold air was dulling as tiny beads of sweat formed on his brow while others tumbled from his stunted sideburns. He looked toward the peninsula, close to a mile away, and soaked in the beauty of the bay, gilded by the sun, now a golden ornament hanging in the sky. His hair dampened as the moisture from a light fog collected in it.

It was a big case, undoubtedly the biggest ever in New Bedford history. The locals had requested access to FBI personnel and resources. DeWald, out of the Boston field office, was assigned as the human part of the equation.

At the moment the case made no sense. Sure the perpetrator was sick, in fact outright deranged, but to go through all the trouble, all the complicated and treacherous steps, and why? He needed a profile. He had seen killers do elaborate things with their victims before, but this was off the scale of his experience. He stopped in his tracks to pick up a half-buried glass chard gleaming from the sand. He stuffed it in his pocket with the intention of tossing it in one of the parking lot trash barrels as he left. He started back up, unhappy he had to pause, but someday it would have cut someone, maybe even himself.

A few minutes later he was in rhythm—his legs, arms, and breathing, all synched, operating as a finely tuned machine. The killer knew the sea and knew it well. Under dark, the sea became a separate universe, a vast, open, almost infinite opportunity to kill. The killer used this to his advantage. But he could not possibly be operating alone; too much was to be done in too short a period of time.

DeWald reached the peninsula and stopped by a knee-high boulder. Nearby, a green crab scurried in a crystalline pool of saltwater, left orphaned by the receding tide. He hiked one leg atop the boulder and stretched, and did the same with the other. His elongated shadow stretched the height of the beach up to the edge of the eelgrass. Time to head back. He walked for a short distance and then returned to his jogging. When he reached the three-quarters mark, he shot into a sprint and stuck with it until reaching the parking lot. At arrival, he noted two men standing by a vehicle parked next to his. Great, just what I don't need—my good friends Laykin and Thomas.

Detective Laykin, New Bedford PD, leaned against his car door. His silhouette tall and stout, a lumbering force to be reckoned with when angered and if within range. The glow from the early morning sun compensated for the coolness of his salt and pepper mustache and hair, giving his face color that otherwise would not have been there. Detective Thomas, much younger—raised watching flashy cop shows on television, dressed in a suit movie actors playing detectives in Miami might wear—paced at the front of the car. His cigarette was down to a butt and he had already reached for another. Laykin's cigarette dripped from his pressed lips. He removed it.

"Little cold to be running," Laykin said without changing his expression.

"For a guy from Minnesota, this is heaven," DeWald replied out of breath. "How can I help you fellows?" He began stretching by bending over his legs.

"We've got to go public. The Chief and Mayor are on my case. I need leads. The public probably has something on this psycho."

Detective Thomas stopped his pacing and shot out, "We look like we're sitting on our—"

DeWald stopped him with a curt hand signal, palm flat out. "I don't care what we look like. What I do care about is getting to the bottom of this case."

"A la the public," Laykin said.

"The public? Do you realize how many loonies will step forward taking credit for this? And do you know what that means? It means we investigate every one of them. And guess what? That takes time away from solving this case. Gentlemen, we have information only the killer could know. Do you understand the value in that?"

"What is this, twenty questions?" Thomas asked, looking over to Laykin. Laykin muttered something under his breath.

DeWald thought for a moment and paused. "Look, I know you guys are under pressure. Let's give it a few more days. We'll revisit then, okay?"

A voice crackled from the radio in the Detectives' car. Laykin picked up the receiver. DeWald could only make out half the conversation. Laykin returned the receiver to his car. He looked sick.

"We've got another victim gentlemen…"

13

Chapter II

The spring at the base of the office chair creaked as he leaned over the lower left-hand drawer of his desk. He paused as he looked into the drawer's contents, seemed to contemplate something in particular, and pulled it out, placing it on the desktop. From the desktop he removed an unregulated power supply, needle bar jig, a diamond-tip liner tub with grip, 3-round stainless steel needle, clip cord foot switch, and tattoo machine with a quick change chuck, placing them carefully in the opened drawer. The ink cup holding a tribal black pigment was summarily tossed in a nearby trashcan. The drawer was closed and the chair creaked as he returned upright, clicking on a portable desk lamp and sliding the object he had removed earlier, front and center.

The blue spiral notebook was well worn, but clean from dust and soil. He opened it, a brittle slice of paper with a slick finish and strained black ink, like that of a microfiche printout, was pulled from a pocket on the inside of the cover. He held it below the lamp. A handwritten date of February 10, 1968 was printed at the bottom.

Pregnant Woman Overdoses, One Baby Lives

A pregnant woman believed to have overdosed on LSD was rushed late last night to St. Lukes Hospital. Doctors, after checking vital signs and attempting some initial life saving measures, determined that the woman could not be successfully revived, and turned their attention to her unborn twin babies. The woman was rushed to surgery where the babies were removed. Only one of the twins survived. Preliminary indications for the survivor are favorable. Allegedly the mother was a single woman and according to authorities, the child will eventually go up for state adoption. The name of the mother is being withheld until...

The thin sheet was returned to the pocket. His attention moved to the notebook itself. As he held it in his open palm, the notebook voluntarily opened to page 67, with a date of August 8, 1967:

Sharon is still on my case about the acid. She insists I quit and she thinks it might hurt the twins. I don't really care much I guess. I mean, I should have aborted them like Dan told me to. I'm tired of puking every morning. It's good enough that I'm letting the things live, so they need to let me live, meaning enjoying some acid every once in a while. Dan made a surprise visit from Canada. He says he can only stay a day or two because of the war and all still going on and he doesn't want to get caught. He brought Linda with him, a whore in blonde hair. The only good thing she did was bringing a supply of something called Thalidomide. She swears it will rid me of my morning sickness. She left me a big bottle that she had leftover from her last kid. She says it has been known to cause birth defects, but her kids are okay. I'll take my chances. My first one, Stacey, I think her parents named her, came out fine and I took worse then. These kids ain't worth puking every morning over. My mother took worse and I'm okay. Got a five-dollar tip from some man who looked like a banker. I think its because I let him cop a feel. Sherry spilled hot coffee...

He stopped his reading, his finger over the highlighted name "Stacey." The notebook was closed and placed back in the drawer, and yet before it came to a complete rest, a paperback book with rounded corners, plagued by foxing, was extracted and placed in his lap. Entitled *Moby Dick*, the book, much as its neighbor in the drawer, opened naturally to a page with a passage haloed in faint yellow highlighter.

With his ivory arm frankly thrust forth in welcome, the other captain advanced, and Ahab, putting out his ivory leg, and crossing the ivory arm (like two sword-fish blades) cried out in his walrus way, Aye, aye, hearty! let us shake bones together!—an arm and a leg!—an arm that never can shrink, d'ye see; and a leg that never can run...

He rocked in the chair and patted a bulge in his shirt that appeared below his shoulder and muttered the name Pip. The bulge jerked.

Ahab's abutted limb of white, cold as stone, could not compare with the living ivory on his chest.

Everything I do—all of this (he scanned the room)—I do for you, Pip, my brother. And you plead with me to stop. But I cannot. Pip I love you. You are with me always. I have been cleansed of my foul state and now I must cleanse others.

He lifted a braided cord from his neck, freeing it from Pip, and studied the silver band dangling at its valley. The chair creaked as he stood, stooping over to turn off the desk lamp and place the book back in the drawer, before leaving the room, as silently and stoically as a priest his sanctuary floor.

Bennett turned up the volume on the Weather Channel and returned to the bathroom sink. He splashed warm water on his face and pushed his hands over his cheeks to his forehead in an attempt to wash the sleepiness from his features, the removal of his hands revealed a pair of baggy bloodshot eyes. He dried his face with a threadbare towel—the same one that hung in his bathroom when he lived with his folks while he attended high school—and reached for the shaving cream, pushed a dollop in his hand, and lathered his morning beard. A smile folded the shaving cream in the corners of his mouth as he heard in the background that it would be a warm day. Finally some spring weather. As he picked up his razor and brought it up for a draw across his cheek, the corner of his mouth curiously took an involuntary downturn as if he no longer had control of it. A thick, clean, stream of drool followed, eroding its way through the shaving cream down to his chin, where it thinned considerably, dangling in the air like a floating silk thread from the cradle of a spider web. Half his face felt heavy and numb. Terrifying streams of emotions, from inexplicable sources, ran through him, as did a powerful sense of déjà vu—he had seen himself looking in his mirror, looking just the way he looked now—before. He smelled something gone rotten.

He rushed from the sink and grabbed a paper and pencil from his nightstand in his bedroom and hastily scribbled a note on the paper, not even sure why he did so. He set the pencil down, walked out his front door, and into the street, standing in only his pajama bottoms, his eyes glazed. A woman driver slammed her brakes as she stopped only feet in front of him. He mumbled and stared with blank eyes.

His mind in a panic, subconscious doors swung open allowing the conscious to peer in, revealing memories, the origin of his trauma denuded.

In the memory a bonfire is raging into the night illuminating all people in its vicinity, its rogue light is casting their shadows to the damp earth in jagged, flickering, extrusions, and this while it noisily devours the timbers it engulfs by igniting air pockets with a bark, releasing crumb-like sparks to the vacant October night. The bonfire is churning at the tip of Wilbur Point. Two trucks with kegs on tap are parked at either end of the point, no more than 500 yards apart. A red Dodge pickup truck is backed close to the outer circle of the fire. In its bed are two towering loudspeakers blasting rock music in the area and across the bay, a high level of wattage driving the bass, causing the ground to rumble. The participants dance on the rock, a large horizontal rounded one at the very tip of the Point, they dance on the circular gravel road around the perimeter of the Point, picnic tables, everywhere and anywhere. Some wear costumes, nothing extravagant, while others wear everyday clothes. It is Halloween night.

Bennett is a sophomore in high school at the time. He is faintly participating in the festivities by staring at the mature enticements, but not knowing how to become a part of it. He is holding a beer, its suds flat; he has taken no more than a few swigs in the last hour or so. Close by, two guys chewing Skoal and wearing cowboy hats are arguing over whom has the better pickup, Ford or Chevy. All around, movement prevails as youths bustling with supercharged hormones attempt to defuse their urges into the night air. Bennett is by himself; seated at the same picnic table his cousin first released him to a few hours earlier. All day he debated whether he should go or not. The party is a tradition at the high school—the first big one of the year, the one where sophomores are obligated to molt their prepubescent skin. He decided to go only after Patrick had persuaded him, or more accurately extorted him, by explaining the high school caste system and how he would be considered a peasant for the duration of his high school days should he choose not to attend. Bennett followed the senior advice and chose to attend.

His bottom aching from its long stay on the picnic table bench, which grows colder as the night wears on, Bennett is up from the picnic table and walking over to the gravel road and begins following it around in order to reach the rock at the tip of the Point. A wave of smoke from the fire breaks on him, and he breaths in the smoke, which burns his throat causing him to cough. His eyes burn as well. He closes them for a moment, but the stinging remains. He hears screaming and laughing coming from the water. Idiots are swimming in the ice-cold choppy seas. He opens his eyes and looks toward the rock, hovering over the granite mass like fireflies in suspended animation, are the burning tips of cigarettes, filled with tobacco or marijuana. Bennett knows Patrick is at the end of one of them. He clambers over the initial rocks, those at the pedestal of the big rock, and almost slips.

"That's gotta be my cousin. He slipped on a rock when he was a kid. Good thing I was there, I saved his freaking life." Laughter follows.

Bennett looks to his left to determine where the words are coming from. He finds Patrick and a couple of other guys. He cannot make out faces, but Patrick's tall thin frame is obvious.

"Is he retarded?" a figure says, and begins choking as he takes a heavy drag from a marijuana cigarette the three are sharing.

"He's just straight," Patrick replies.

"You mean like a geek or something?" The same figure asks.

Patrick takes a long drag and flicks the roach to the ground. "Yeah, or something." He walks up to Bennett. Despite the darkness, he can sense the lost look in Bennett's eyes. "Are you ready to go?"

Bennett does not hesitate with his reply. "I think I've experienced enough."

Patrick chuckles. "Alright man, let me get one last brew for the road and I'll meet you back here."

Bennett watches as Patrick scrambles over the rocks to the gravel road and in the direction of the beer truck at the left end. Bennett pulls his arms up across his chest.

"—You a computer nerd?" the figure asks. A joint burns from his hand, resting flat against his leg.

"Yeah, I guess you could say that," Bennett replies. He continues to watch Patrick, now two people back, in the line to the tap.

"How good are you?"

"Good enough." Bennett thinks about it and continues, "I can hold my own."

The figure takes three hits on the cigarette before speaking again. "I've got a 20 megabyte hard drive in the trunk of my car."

Bennett makes an auto response. "Cool."

"You want it?"

"What, for free?" Bennett replies, now more attentive.

"Yeah right. I'm asking seventy. Want to take a look at it?"

Bennett looks for Patrick. He is stooped over at the tap. "Patrick should be back any second. I'd better not."

"It will only take a few seconds. My car is right over there."

Bennett looks in the direction the figure is pointing. He can see a car, but little more, it is over in a corner away from the bonfire. His eyes glance over to a girl dancing on a picnic table with a crowd of guys around her, whooping and hollering, urging her to strip.

"Are you gonna check this out or not?" the figure asks, impatiently.

"Alright," Bennett relents, taking one last peek at the girl to see if she is being responsive, but she does not appear interested.

As the two make their way to the car, the man reaches out his hand towards Bennett. "They call me the Cap'n," he says.

"Is that for Captain?" Bennett asks, shaking his hand.

"No, it's for General."

The sarcasm is not lost on Bennett. "What are you Captain of?"

The man places the roach in the side of his mouth and raises his arms out straight. "All this…" he mumbles, rotating his shoulders to cover the area.

"What's *this*?"

The Cap'n takes a long drag and speaks as he exhales a cloud into the crisp night air. "Space, man. It's not here yet, but it will be, mark my words…it will be."

"What kind of space?"

The Cap'n chuckles, knowingly. "An electronic dimension; a place where reality and the imaginable, coalesce into a single realm, whereby it becomes impossible to distinguish between the two—a unification of the senses and the mind's eye; an extensible median removing time and geographical separation, illuminating freedom and fetters, love and sin, crime and consequences. And best of all," he said as he turned his self-absorbed face in the direction of Bennett, "you can do all this without leaving your house." He takes a drag. "I'm talking FidoNet, I'm talking MUD, I'm talking the next generation."

Bennett knows the pot is partially responsible for the man's existential exposé, but some other chemical, or lack there of, is in the blend too. He notices a pirate's accent moving in and out of the man's speech pattern as he talks, scaling with his excitement.

The latch on the car trunk is released and the trunk lid springs up. The man leans in and retrieves a thick rectangular object.

"Here you go lad," he says, holding the object in front of Bennett. "Quiet, wench!" he yells at some elderly woman who is on her porch and is vociferously complaining about the noise.

Bennett reaches out for it. It is difficult to judge distance in the darkness, leaving Bennett a challenging task to grasp it. The Cap'n causes the object to jump when Bennett is attempting to clutch his hand around it. He lets loose a hearty pirate's laugh.

Bennett withdraws his hand and shakes his head. "Forget it. I've got to go—"

"I thought you enjoyed playing games."

"Not this type, besides, how would you know?"

"How would I know? Is that what you asked? How would I know? Well, since the element of surprise is a part of just about every game—you'll just have to be surprised that I know. And speaking of surprises, I have another one for you!" His laughter roars.

It stops. And after a second of silence, the Cap'n draws back and smashes the object against Bennett's head with a dull thud. All is black.

The memory faded, and the door swung shut, once again closeting the subconscious, frightened thought. Bennett found himself in the street, wondering how he got there, and was startled by the honk of a car only a few feet from him. He groggily moved to his yard and in doing so discovered his muscles felt sore and stiff. As his mind began lumbering to a conscious state, nausea swept through his stomach and as he walked back inside the house, he heard in the background the TV blaring about the weather. He felt something clinging to his face and reached his hand up to rub it. A gooey substance coated his cheeks. He panicked and began scraping it off with his hands. A very familiar odor flowed from the substance. He brought it close to his nose. Shaving cream. The nausea took hold of him and he started the dreaded journey to the bathroom. On his way he found a piece of paper with a note in his own handwriting.

"You've had a seizure. You'll be okay."

They made an odd team, DeWald thought, on his drive to the crime scene as he contemplated the group dynamics between Laykin and Thomas and himself. Laykin, a hardened veteran with a touch of the old school on policing, known to be a hothead and impulsive investigator, married to the same woman for forty years. Thomas, a young kid not long from the doors of the academy, but far enough to enter the door marked detective; called Junior by most of his peers because he was considered sired by Laykin, and stretching his leash to make his mark, and the flashier the mark, the better. And himself, a few years with the Boston field office, his footprints still lingering at Hogan's Alley in Quantico, and yet a considerable number of cases under his belt, eager to perform his job in a meticulous and factual manner operating with a high degree of self-discipline and scientific method. Laykin had already labeled him publicly as the General Bernard Montgomery of law enforcement.

He drove alongside a parade of law enforcement vehicles placed against a backdrop of heavily wooded and undeveloped land. Deep forestry for such a narrow peninsula, he thought. The vehicles represented Patrolmen, Detectives, the County Coroner, the office of the DA, the crime scene search unit, and his least favorite—the press. At the end of the parade, DeWald made a U-turn and pulled behind the last car.

Gingerbread cottages lined the opposite side of the road. The elderly residents were gathered in their front yards in blue-crested clans. A man broke loose from a circle and made his way in DeWald's direction, his nose covered in white sunscreen and his eyes with big square wraparound sunglasses; he looked determined. He demanded to know what was going on and said his request spoke for the entire group. DeWald said hello and informed the man that it was official police business and he could not discuss it with him now. He did let the man know that nobody was in any sort of immediate danger. The elderly man turned to face the circle of people chatting nearby, yelling back to them that he had discovered another worthless law officer. He waved off DeWald and turned to walk back to the circle.

"What a circus," DeWald said under his breath. He looked down the road and saw Laykin and Thomas. Laykin was pointing at him, and both he and Thomas were laughing. Well at least they're in somewhat good spirits, he thought. Looking up he saw ultramarine skies occupied with white cotton candy clouds. His stomach growled. The sky was in stark contrast with the ground, a bustling of flashing lights and hurried personnel darting around the cars. When he reached Laykin and Thomas the smiles had dropped.

"So what have we got?" DeWald asked, looking at the end of the road, the only direction where a beach was in sight.

"Wrong direction, DeWald," Laykin spoke up. "We've got a walk ahead of us—"

"Through the woods to the other side of the peninsula," Thomas chimed in. "We cut through the woods." He took a quick succession of hits on his young cigarette, tossed it down, and drilled it into the earth with the tip of his shoe. "What a waste. Told yah he'd get here before I could get very far on my cigarette." He looked at Laykin who mimed a violinist.

DeWald observed a man in shorts and a T-shirt talking to a police captain. "Great, the press will be getting in our way."

Laykin dropped his cigarette to the road and crushed it beneath his shoe. "The hell they will. They don't go beyond this point until we've taken care of the crime scene."

Thomas headed towards a tight clearing. DeWald could make out others in front of him, but they were nearly invisible through the thick brush.

"Very similar to the other ones," Laykin said, out of breath, as the three began down the trail.

"Not quite," Thomas said.

DeWald studied the ground as he walked. "How so?"

"The victim is a celebrity of sorts," Laykin said.

Thomas released a branch that he had pulled out of his way and it thrashed into DeWald's chest.

"Hey watch it there, Junior" DeWald barked. "Who is the celebrity?"

"Kathy Rueger...at least we think it's her," Laykin said.

"Trust me, it's her," Thomas said.

"Oh that's right, you've got that photo of her in your locker—"

"Yeah, took it from you."

"Celebrity killing?" DeWald questioned silently. The first two victims were everyday women—one, a freshman at the University of Massachusetts, the other, a barmaid at a local tavern. Was the killer moving upscale? Or was Rueger in the wrong place at the wrong time? Given the first two had nothing significant in common—at least the evidence uncovered had indicated as such—the odds were Rueger would not have anything in common with them. But this was still to be determined. DeWald had worked similar cases before where on the surface there appeared no apparent pattern to the victims, but in the end, invariably, the pattern was there, once it had been dissected from the psyche of the killer, the only amenable host to such detestable thoughts.

The heavy forest gave way to dense shrub, which sequestered a brackish looking pond, painted turtles sunning themselves on a decayed log floating in its center. Twigs snapped in the woods at a distance, indicating to DeWald that officers were performing a warm search, a search beyond the immediate scene of the crime. Word made its way down the train of investigators to watch out for poison ivy.

"You know what poison ivy looks like, Agent DeWald?" Laykin yelled out once the word went beyond DeWald.

"—That, *and* poison sumac *and* poison oak."

"Got ourselves a real Boy Scout," Thomas remarked.

"That's Eagle Scout to you, Junior," Laykin said sarcastically.

"Wrong," DeWald said. "Biology major."

"Then you should be leading the way," Laykin said.

"Too late for that."

"Why is that?" Thomas asked.

"That brush you two just marched through…"

"Yeah?"

"…Poison sumac," DeWald said.

Both Laykin and Thomas stopped in their tracks.

"You're bulling me," Laykin yelled.

DeWald shrugged and smiled big. "So don't believe me. But I'd suggest you two wash thoroughly when we get to the beach."

The group broke through at the other side of the forest, and both Laykin and Thomas bolted for the water, and as they scrubbed their hands viciously in the surf, DeWald chuckled, wondering what poison sumac really looked like. He walked over to the white bag containing Rueger's lifeless body and motioned for the medical examiner to open it up for him.

DeWald's hand involuntarily grasped his nose and mouth. "Whew." He inched forward and sniffed. "What a strange mix of odors. What am I smelling?"

"Not sure, but a large part is probably coming from decayed krill," the examiner said.

"Krill?"

"Thought you were a biology major?" Laykin asked, huffing from his walk up the beach to the crime scene and wringing his hands of the saltwater.

"So?"

"They're tiny shrimp. The whale's baleen filters them out. It's what he eats."

DeWald nodded. "Must have missed that day in class." He hesitated for a minute. "So what's baleen?"

Laykin and Thomas looked at each other. "You were screwing with us about that Biology major weren't you," Thomas barked out.

DeWald grinned. "Maybe."

Thomas and Laykin shook their heads in unison. DeWald watched as Laykin's brows shot up and down in rhythmic spasms and this accompanied by his eyes agitating in their sockets. It was something Laykin had a habit of doing, inexplicable and involuntary, but annoying just the same for DeWald.

"So what's the answer to the question?" DeWald asked.

"Whalebone," Laykin spoke up, down to one brow flinching.

"He's the only one old enough to remember the term," Thomas joked.

Laykin gave a before-I-was-so-rudely-interrupted look and continued. "I read a good article on this once—"

"Ah, yes, the Yankee in the men's room," Thomas said.

"Junior, kiss my—"

DeWald jumped in. "Get on with it, Laykin."

Laykin rubbed his eyes. "According to the article it was called whalebone in the early days in New Bedford. It's some sort of hairy crap that hangs from plates in a whale's mouth. It's like your fingernails, though stronger. They built buggy whips, corsets, umbrellas, and stuff like that with it. Once was more valuable than blubber."

"Worth more than your chest, huh?" Thomas took two steps back.

"No, that would be your brain," Laykin said.

"Don't ask him about black holes or quantum physics," Thomas said.

"Just a hobby." Laykin cracked a smile.

"Thanks for the History lesson, professor," DeWald said.

23

DeWald and Laykin moved closer to the body bag. The strong scent of perfume wafted up at both men causing Laykin to take an exaggerated sniff. He motioned to the examiner.

"Smells like she put a whole bottle on herself," Laykin said.

"You got samples of this perfume?" DeWald asked the examiner.

"Yes, sir."

"Can the lab identify a brand?"

"I'm no expert, but they should be able to. I think just about every perfume out there has a unique and patented formula—"

"Hell, it's Chanel No. 5," Laykin blurted out.

"How do you know?" DeWald asked.

"It's all Delores ever wears. When you've been married to a woman forty years, you know her perfume, trust me."

"Check it anyway. I don't trust your nose, Laykin."

Laykin shrugged. "What's that thing on her forehead?"

DeWald, donning gloves, bent over and gently pulled the crusty strands of hair partially covering her forehead, to one side. "Looks like a tattoo. Same as the others."

"Same tattoo?"

"Yeah. If I'm not mistaken, it's a picture of a whale."

Laykin bent down and joined DeWald. "Sperm whale I'd say. Or, I'm sorry, what's its name in Latin, Mr. Biology major?"

DeWald rolled his eyes and looked-up at the examiner. "Will you be able to tell how long this woman has had this tattoo?"

The examiner shrugged. "Probably. If it's recent, I should be able to spot perturbation to the surrounding skin."

"Check that for me—"

"She was on the six o'clock news for crissakes," Laykin shot out, not taking his eyes from the body.

"Have you seen the makeup they wear?" DeWald fired back.

"Are you saying she could have covered this up?"

"She could have covered a hole in her head."

"She could have never removed it then."

"How so?"

"She was a bubble head…think about it."

Thomas laughed from the background. "Sweet."

DeWald shook off the remark and returned to his study of the body. Her skin was spectral. She was fully clothed. The other two had been the same. He thought about Rueger. What a contrast. From the beautiful, vibrant woman espousing her opinions across the airwaves to thousands of households—to a grotesque, motionless, silent victim, whose entire outreach was limited to the three inches between her body and the bag—all this in one night. The red ring around her neck and blood clouded eyes told DeWald all he needed to know. She most likely died by strangulation, just like the other two.

DeWald surveyed the crime scene. The wealth of the ocean fell to the perpetrator's side, its caustic reach scouring trace evidence from the whale. The sand beyond its claim had been raked by brush, smooth as a trap at Pebble Beach. Something near the whale caught his eye. A large chunk of gray material, lathered with busy horseflies, lay not far from the whale's corpse. "What's that thing?"

The examiner looked in the direction. "Its tongue."

DeWald exaggerated his stare. "Tongue?"

The examiner shrugged. "Yeah, they had to remove it to fit the body in there."

"I haven't seen a tongue at the other crime scenes."

"They probably got a little sloppy. Maybe they were hurried."

"Even with the tongue out it looks like it would be a tight fit."

"Humpbacks have throat groves on the underside of their lower jaw. The grooves expand like an accordion, forming a pouch. It allows them to take in huge volumes of water to strain food from."

A tiny yellow plastic flag hung docile from its wiry pole, stuck in the sandy beach up near the tree where the rope had been tied, an Arctic white blotch, settled in the soil like a smashed snowball, at its base. DeWald thought of this spot, and the equivalent ones he had encountered at previous scenes, as that singular dapple of purity in an otherwise grotesque display of vileness. It seemed so odd, so out of place, something that didn't belong there, and yet there it was, at the base of the tree at each and every crime scene. He knew what he would find if he were to lift the white dental stone object from the sand and turn it over to look at what was underneath. It would be no different from the others; it would yield the same bump with the same five stubby appendages. It would bear the small porcelain-like hand of a child. Tiny and fine like the hand broken from an angel statue resting in an herb garden, but unlike the variegated grays that colored a garden statue, it manifested a brilliant Artic white color, pure as snow from the Poles. The forensic specialists had expelled the print from the earth into the dental stone, but who had put the impression there in the first place?

The car window down, warm spring air swirling, commingling with a rock tune on the radio, Bennett turned down Johnny Cake Hill and pulled in front of the prestigious New Bedford Whaling Museum. He paused in the car, still enduring faint symptoms from his seizure, though most had dissipated. Before he left, he placed a sign on the back of his front door:

"Take medicine!" Out the passenger window, not far down the Acushnet River, the Island Ferry pulled into dock as cars piled into the Crab House restaurant. Bennett could not recall the last time he had missed a Society meeting and he was looking forward to this evening's, intrigued by how the Society would respond to the disastrous interview with Kathy Rueger. He recognized Collin's pickup, Will's Toyota, and Adam's Dodge.

He dashed up the front stairs, entering the museum and its rich historical ambiance, and when inside, greeted by a full-sized whale skeleton strung from the towering arched ceiling near the far wall, where in the darkness the yellowed bones loomed larger then life, eerily floating in time and space, ignorant of nature's prime directive to recycle. Below, the bark Lagoda, fastened within the wooden floor, a white band stretched around its near black hull, highlighting its below deck portals, while above, its yellowed sails hanging from its intricate rigging. On the wall near Bennett, spanning its entire length, an ivory Moby Dick, captured in oils, suspended in sea green, his brown brothers swimming along side in the background.

He took a quick turn left down a narrow hallway to get to the research library where the meetings were held. The building wore its 1907 construction well, the structure exuding an unmistakable historical, yet fresh scent. The ubiquitous paintings of turgid waters, whalers in a boat abreast a whale, the harpooner at the bow with harpoon in hand ready to strike, and the main vessel in the distant background with seagulls peppering the skies—reached Bennett in a way he could not explain.

Maybe it was his knowledge of an ancestor, a harpooner named Jack Howland, who in 1841, as the story went, set sail aboard the *Yankee* out of New Bedford. On that voyage, the *Yankee* encountered two other whaling ships off the east coast of Japan and the three converged on a tremendously large sperm whale with a notorious reputation for sinking whaling ships. After sounding, the rogue whale burst into the air between the whaling boats from all three vessels. Jack and his boat were near, allowing him to secure his harpoon into the blubber of the beast. In the end, the beast won, crushing Jack's boat and all the men in it, turning the waters crimson. Jack died that day, leaving his family behind. Reportedly, the whale went on to carry many other men to their graves, his name sending shivers down the spine of whalers, and serving as the impetus for a writer named Melville to write his novel about a fictitious whale with a similar name. Vessels were likely to avoid him, once his name rang out from the crow's nest: "Ahoy! Mocha Dick."

Bennett stopped at a glass display case embedded in the hall wall. The lights were out, but the outline of a sperm whale's tooth, that in the light would depict intricate scrimshaw of the Uncas whaling off the Cape of

Good Hope, was unmistakable. He had the artwork memorized and could make it out in the darkness. The door to the restroom opened up behind him.

"Staring at that dad gum tooth!"

Bennett jumped and spun around to see Collin shaking his head.

"Damn, Collin, don't scare me like that."

"The bar…the other night. Remember? Payback," Collin said with satisfaction.

"Touché."

"You know it's only a rumor that Jack Howland carved that tooth."

"Scrimshaw, Collin," Bennett corrected.

"Whatever. What's the fascination—it's just a dad gum tooth."

Bennett thought for a moment. "I'm not sure. Maybe it's because these guys had audacity. You've got to respect that. They're the last of their kind."

"All that from a tooth?"

Bennett tapped his finger to the glass, pointing at the tooth. "It sums it up. That tooth is saying 'I'm not afraid of anything.'"

"The whales are our friends now, why fear them?" Collin said in an effeminate voice and wearing a crafty smile.

Bennett feigned a cough that formed the word *moron*.

Collin pulled a sucker punch at Bennett's stomach. "Chill-out." Bennett didn't change his expression. "Guess what?" Collin didn't wait for an answer. "We have a babe watch on tonight. Second row from the front on the left end." Collin shifted his eyebrows up and down.

Bennett's stoic expression broke. "Blonde or brunette?"

"Neither…" Collin replied, lavishing his answer. "Redhead."

Bennett bobbed his head. "Lead the way…"

Eight rows of chairs were centered on the floor of the research library. John Phillips stood at a podium placed at the head of the rows. As usual, the artsy side of him had colorfully filled a white board with the meeting's agenda and upcoming events. Bennett scanned the board and then looked for the redhead. He would have scouted the audience anyway, as was his habit. He kidded himself, thinking he never knew when a voluptuous woman might take a hard knock to the head, fall unconscious, and awake with an inexorable urge to attend a NBHWS gathering. Each night it could happen, however remote, it was possible. Perhaps tonight it had really happened. He slid into a seat in the row behind her, directly lined-up, detecting a fresh scent, perhaps freesia, wafting from her hair. Intoxicating. He wished she would turn around, and hoped when she did, his imagination of her beauty, when placed against her physical features, would become indistinguishable. He tried to picture her face. He was sure she was beautiful…

"Okay, let's get started," John said from the podium. He pointed to his flamboyant artwork on the white board and began rattling off upcoming events.

Bennett felt a flush in his cheeks. He hated to admit it, and even felt guilty over thinking so, but he had to acknowledge his embarrassment over the geek tone of the meeting and how he was certain the girl in front of him would perceive it that way.

"...I see some new faces in the crowd," John continued. "We don't like to embarrass new folks, but we do like to provide a warm welcome. If you are new today, if you wouldn't mind, could you stand up and briefly introduce yourself so we can get to know you better after the meeting..."

A balding man, his head mottled with liver spots, wearing thick clear, but yellowed, plastic rimmed glasses, rose from his chair. "Alfred Huffnagle," he began, his fingers fondling his sparse white mustache, which appeared to consist more of milk residue than hair. "Me and the misses just drove up from Florida to stay for the summer. She plays bridge on this night so I needed something to do. I heard about you guys on the news. Is it—?"

"Thanks, Alfred, and welcome," John said.

Alfred lowered himself back into the chair muttering under his breath. John looked around the room for others—the redhead jumped-up from her seat in a single spunky motion.

"I'm Tanya Appleton," she said, while at the same time attempting to catch her breath. She turned to face the group behind her. Bennett's heart palpitated at her striking looks. Poised like emerald coronations atop her high and chiseled cheekbones, her green eyes illuminated in stark contrast with her auburn hair. She wore little makeup, and probably did not require what she had. Her pink gloss lipstick brought a vivacious spirit to her taut lips.

"I'm new to the area," she continued. "I visited the museum earlier in the week and was intrigued. I saw your announcement on the bulletin board so I thought I'd check it out. You know, meet some people."

A moment of silence followed her statement before John realized she was finished and he needed to say something.

"Ah...welcome, Tanya," he sputtered.

Bennett thinking, she's beautiful.

As John finished the greetings and moved down the agenda, Bennett continued ogling the back of Tanya's head and breathing-in her freesia perfume.

"...I think we should make a public statement condemning the harpooning of that whale..."

John's last statement jolted Bennett from his reverie. He was amazed that subconsciously he had paid enough attention to hear it. The statement irritated him and he found himself speaking before he had completely broken free from a daydream of Tanya.

"Why?" Bennett called out.

John looked caught off guard. Bennett peered over at Collin, looking for moral support in Collin's expression. Collin looked away.

"Why not?" John said. "We're all against it, what's the harm?"

"It's the principle of the matter," Bennett replied. "We haven't done anything wrong. We know whoever killed that whale has no affiliation with our organization. Why should we make a statement? I think it makes us look like we're caving. If we ignore her, I think it makes what John said on the show more substantive."

"Sorry, Bubba," Collin spoke up, "but I agree with John. If we make a strong statement, I think the public will be more inclined to back us. Right now we're practically a secret society to the public; they may think the worst if we don't try to educate them on where we stand on the killing."

Bennett sighed and gave Collin a look of disgust. He noted Tanya staring at him intently. He wondered what she was thinking. He hated to admit it, but the uncertainty was urging him to back off, to cut his potential losses. Principles? Hell, he obviously had none.

"Well, what do you think, Bennett?" John asked, thinking Bennett was contemplating the issue at hand.

"Umm…" Bennett tried to say convincingly, as if he were thinking it through. "Let's vote on it." Just about everyone in the group either shrugged or nodded. He sighed, relieved an escape idea had come to him before he had to say more.

Alfred fought his way out of his seat. "Sounds like a heckuva good idea to me. The wife's bridge game is almost over and I need to skidattle."

The room broke out in laughter. Alfred smiled and sat back down.

John nodded. "Let's vote," he said. "All in favor…"

At first the audience fell to considerable head turning, but in time a few hands shot-up, including Collin's, and then the remainder of hands went up, excluding Bennett, and much to his surprise, Tanya.

"I'd say the motion carries," John said, and in jest slapped his hand on the podium.

A few minutes later John adjourned the meeting. Alfred whisked up to John and shot into a conversation with him. Bennett took a deep breath and tapped Tanya on the shoulder.

"Thanks for voting with me," he said.

Tanya turned around and greeted him with a big smile. "No problem. I happen to agree with you."

Bennett felt an awkward moment approaching as his mind fumbled for something more to say. He felt his cheeks begin to burn.

Tanya grinned. "I think we deserve a cup of coffee, you know what I'm saying."

The thick smoke flowing from the next table—as a burly looking man in a bushy beard and stained coveralls, tapped his cigarette against the edge of an orange plastic ash tray—caused Patrick to cough, and after adjusting himself on the brown plastic bench of his booth, he ran his finger across the art deco table top and observed the trail left behind in a film of grease. He motioned to the waitress to refill his bottomless cup of coffee, and looked around the café called *The Trawler*. It was no misnomer; every man in there made his living fishing and every man in there reeked of the peculiar odors of his occupation. The stern forces of nature carved their way uniquely into each man's face. Patrick knew them better by the crevices in their skin then by the color of their hair. The *Lock and Load* had pulled into port that day after a three-week stint. The men were getting their fill of beer and Bruin's hockey. Patrick rearranged again on the plastic seat in the booth and emitted an awkward sound.

"Hey, fart outside, man," Paul said, as he slipped into the other side of the booth.

"Where've you been?" Patrick asked, perturbed.

"Working. I've got a new design. It's great—this puppy is sticking to its target come hell or high water."

"Listen to the MIT grad."

"It's a gift, man. If Ahab had one of my harpoons—Moby Dick—dead! You hear me—dead!"

"Wise up. It's just a stinking job."

Paul shook his head. "You don't get it, do you? It's a *primordal* thing—"

"You mean primordial?"

"Whatever. I get such a rush jamming my tool into the back of those whales—"

"Sounds more Freudian then primordial—"

"Would you quit interrupting me. Look, I'm just saying I'm enjoying this—it's more than a job to me."

Patrick noted the serious expression on Paul's face. "Well it's a job to me and we're supposed to get paid today. I need you here when we get paid. So settle down, Paul, settle down."

Paul gave Patrick a blank stare.

"This guy is a loon—a member of some weird cult or something. The guy wants us to call him Ahab—think about it. Hell, I don't know what he's doing with these whales, but it can't be good."

"Maybe he's eating them."

"Eating them? Who eats whale blubber?"

Paul shrugged. "Eskimos?"

"Does this guy look like an Eskimo to you?"

Paul thought for a moment. "No."

A holler broke out from the crowd at the bar.

"I see *Lock and Load* is in port," Paul said, looking over at the rowdy group. "Boy, that was close the other night, they almost saw us." He chuckled.

"It's not funny. That's what I was saying earlier—you're getting so excited about all this crap that you're taking chances."

Paul flicked a toothpick from the table out onto the floor. "Look, man, if I go, I'm going down in flames, none of this silent crap for me."

"Well, enjoy the solo ride down."

Paul shrugged while looking off in the direction of the door. Patrick turned to see the customer walk in. The customer motioned for Paul to slide over and took a seat next to him.

The man's saggy bulldog frown went stern. "You're getting sloppy. I don't like sloppy."

Paul reacted. "Who do—?"

Patrick put his hand up in Paul's face. "What are you talking about?"

The man answered rapidly without changing his expression. "First you leave one bobbing out in Buzzards Bay and now you leave a tongue on the beach."

"Who cares—"

"Paul, shut up," Patrick said.

Paul shook his head and pushed air through his lips, leaning as far back against the booth as possible.

"What's the problem?" Patrick continued.

The man motioned the waitress for coffee. She set down a mug and filled it to the rim. He poured four bags of sugar in the cup. "It's not clean." The coffee spilled over down the sides of the cup.

"I didn't realize clean was part of this deal," Patrick replied.

"—The Bruins suck!" Was yelled from the bar.

"I don't care what you do or don't realize, I'm telling you now this is a sacred thing we're doing. Clean is important."

Patrick chuckled under his breath.

"You find this funny?"

"Hey, it's none of my business, right?"

"Damn straight."

"You got the money?"

"Of course. But first, just to make sure we understand each other. You leave a tongue, tail, testicles or anything else laying on the beach like that again and it comes out of your pay."

"You don't need to worry about the testicles, we only kill cows, as you requested."

"Look, don't screw with me. I can make both of your lives a living hell with nothing more than these." He held up both hands, palms turned toward himself.

Patrick smirked. "Let me guess—registered weapons?"

"They say the pen is mightier than the sword, I say the keyboard is mightier than the pen. I can make you disappear like that." He snapped his fingers. He took a big swig of the coffee, swallowed it, and laughed so that his chest heaved as he absorbed the puzzled expression on both Patrick's and Paul's faces. A wad of bills was handed off to Patrick. The man took a last gulp and darted from the café.

"Who the hell does he think he is?" Paul shouted.

Patrick watched as the man disappeared around the corner of the cafe. "Someone to watch very carefully."

Paul looked on. "Why?"

"Ahab is nuts."

Chapter III

The morning paper blasted against the front door, but went unnoticed. Bennett sat down in front of his laptop computer, his actions detached from his thoughts. He could not stop last night's events from playing out in his mind, and any detour he attempted with his thinking led to Tanya's delicate features as he observed them while seated across from her in a booth at the Uncas, as they talked into the late hours of the evening. Although signs of caution flashed by as he progressed with his thoughts, he continued to ignore them, choosing to allow the threads of his thinking to form emotional ties with a woman he had known but a few short hours. As they had walked from the Uncas, as Bennett's heart pounded through his shirt and his palms turned to the basin of a rain forest, he asked if he might have lunch with her, soon, if possible. She smiled and said she thought that would be nice. As he sat in his desk chair, he watched in his mind's eye, last night as she stepped into her car and drove off, a mundane set of actions, performed so magically.

He shook his head and forced himself to insert the floppy containing the file *Wolf Spider* had trapped, into his laptop. The day before he couldn't wait to delve into an analysis of the file, now it seemed it could wait, but it wouldn't, as somehow he managed to turn his focus from auburn to the black and white on the laptop's screen.

He listed the floppy's contents. The lone *whale* file still lingered in the listing. He issued a command that gleaned human readable information from the code. Like a biologist dissecting the internals of an animal and trying to determine how the organs inside worked together to form a complete system, he examined lines of text, internally stored within the *whale* file. Aside from the information stored in every program for his

33

particular flavor of Unix, the text contained no intrinsic tendencies, and worse, in spots exhibited signs of mangling, perhaps encryption.

He pushed the laptop back and thoughts of Tanya drifted in. He ran his hands through his hair, taking a deep breath and attempting to refocus, trying to convince himself that unraveling the mystery behind the file would somehow prove rewarding. He smacked his lips and pulled the laptop back. Time for step two: run the file through a decompiler. The decompiler would take the gibberish contents of the *whale* file, something only the computer operating system understood, and attempt to reverse engineer it back to the human readable code that it was generated from in the first place. He set some things up and issued the command and waited as his laptop crunched away on its task. He cracked open the window behind his laptop and allowed the spring air to circulate.

The hard drive on the laptop continued to whine. He leaned back in the chair and looked around his home office. He glanced at a wall where prints of whaling paintings hung, the ones he purchased from the whaling museum gift shop and later framed. His eyes returned to his desk where he clutched a scrimshawed whale's tooth—a phony one. He ran his fingers over its intricate crevices. Amazing. A gust of wind burst through the cracked window and spilled an unstable stack of computer periodicals from the coffee table behind him to the floor, while sprawling one copy over his brown dingy sofa.

The latest newsletter from the Woods Hole Oceanographic Institution or WHOI, fluttered to his shoes. The newsletter reminded him of the volunteer work he did for the institute a year earlier. During the stint he had designed and developed their archives photo database—a database that provided access to the institutes wealth of images. He had since thought about the software and contemplated how to make it faster, perhaps utilizing in-memory databases. He really enjoyed his time at WHOI—there was something in the ambience there that pulled him in. Maybe one day he might decide to leave his impersonal silicon companions and congregate near the warmth of living creatures.

Bennett picked up the newsletter and magazines and narrowed the crack in the window. He looked to the street below—he always had a good view from the second story. The hard drive stopped its whining.

"Let's take a look at your genetic code," Bennett said aloud. He brought the output from the decompiler up in a text editor and began slowly scrolling through the results. He studied the top of the file.

```
#include      <stdio.h>
#include      <string.h>

char author[] = "Morye Malchik";
```

Morye Malchik?—The perpetrator, or just the developer? He made a note to do searches on the name. Everything looked normal until he encountered a very large string, or text, that appeared to be encrypted. Bennett shook his head. As he scrolled further, he encountered additional disappointment—the decompiler was confused at some junctures in the code and delivered garbage. He realized he might have to run everything through a debugger, a tool that would represent the instructions in the file from a position of compromise, somewhere in between human readable code and machine code—not pretty and extremely tedious, the ratio of lines of code to an operation just shot up—not for the faint of heart. He got up from his chair and headed for the kitchen and put on a pot of coffee. This was no two-hour job. The owner of the code had ensured its deciphering would be a major undertaking. Bennett sensed a challenge. The dual was on. He chuckled, wishing he still possessed the thinking capacity God had blessed him with at birth, knowing with it, he could solve this in no time.

DeWald walked through a hallway with a high-gloss floor to the Mayor's office. He discovered the Mayor—a thin elderly woman wearing glasses with a red frame and small oval lenses—seated behind a cherry wood desk and in her hand, a series of documents fanned out as if she were holding cards. The Chief of Police—a plump blue Weeble in full uniform, including his chiefly cap—sat on the other side of the mayor's desk, joined by Laykin and Thomas. Silence and no eye contact greeted Dewald—something that led him to believe he was the topic of discussion. The Mayor wore a grim expression. She looked as though sleep kept its distance from her the night before. Thomas had his pen tapping against a legal pad, his knee bouncing to the rhythm. The Chief worked a piece of gum, his jaws operating almost in slow motion. Laykin looked out the office window. The Mayor spoke up and introduced herself and the Chief, to DeWald. DeWald took a seat next to Thomas.

"Gentlemen," the mayor began, her eyes focused on the papers in her hand. "I'd like to take a moment to summarize some of the paperwork that has crossed my desk recently, and I'd like to do so without interruption." Her eyes left the papers temporarily to focus on Laykin. "I have a document here from *Greenpeace*. It's a letter informing me that their flagship, the *Artic Sunrise*, has weighed anchor in Australia, and set course for New Bedford, Massachusetts." Her eyes left the documents and moved across her audience. "Next, I have a series of letters from state environmental agencies—no need to waste our time here. These are followed by what I've categorized as threatening letters. One in particular states that justice would be served our whale killers by having them publicly quartered—quartered by whale, not horse—and this is one of the less painful punishments suggested. After that I have an article I printed from the Washington Post's web site. It's entitled, and I quote, New Bedford: Has Whaling Returned? In the body of the article the writer questions whether our city has gone the way of the Makah Indians." The papers were consolidated into a single stack and placed neatly in a corner of the Mayor's desk. "Gentlemen, all hell is about to break loose—"

Laykin snapped, his eyebrows flowing in waves of spasms. "I don't give a damn about the whales and this whaling business! We've got young women dying here, and this is the kind of attention we get! If you're suggesting we expend resources to take care of our whalers, then you can have my badge on your desk right now! There's little girls lives at stake here..." his eyes welled and tears noticeably streaked down his cheeks, his voice dropped, "...damn the whales..."

The tears, more than the harsh words, appeared to catch the audience off guard, with the exception of Thomas who just as easily might have been lounging in front of a television watching a sporting event.

DeWald cleared his throat. "You know Detective Laykin, the two are not mutually exclusive, we catch our whale killers—we catch our killer."

"Detective Laykin," the Mayor said, authoritatively, "I would request you maintain your composure in my office, or you will no longer find yourself in my office. Is that understood?"

Laykin applied a yellowed handkerchief to his damp eyes and nodded.

"Now," the Mayor continued, "if I had not been interrupted and allowed to present my conclusions on what all these documents mean, I would have stated that a deluge of interference is in route to our nice little town, and that you had better be prepared to deal with it. I thought it best to give you men a heads up. We have to give the individuals who wrote these letters the benefit of the doubt—they are not aware that the whale killings are tied to a serial killer, or that we even have a serial killer." She was looking at DeWald; she turned to face Laykin. "So we can't quite label them as insensitive, can we? Now, quite frankly I don't see what this has to do with how the police force chooses to focus its resources. But the fact is, our city is getting a black eye—deserved or otherwise. Does it hurt business? Yes, of course it does. Tourism? Probably. Obviously the sooner we catch our killer, the better the odds of us heading off tumultuous times for our city."

The mayor looked out the window and sighed. "Now enough said on that subject. How about some status? Special Agent DeWald, where are we on this case?" The Mayor rested her folded hands on her desktop.

DeWald thought for a moment and then said, "This is a very strange case—"

"That's not what I asked." The mayor pursed her thin pink lips.

"Well then, Mayor," DeWald said with a shrug, "we're not as far along as we'd like to be."

"Any suspects?"

"No."

DeWald observed Laykin making eye contact with the Mayor, as if he wanted to see confirmation that he had been right in an earlier discussion on some point. The Mayor shot a glance over to the Chief. The Chief sat up straight, stopped his chewing, and cleared his throat, attempting to sneak in a sigh, but failed. He looked over at Laykin and back to DeWald. "Don't you think its time we involved John Q Public?"

DeWald crossed his legs and cocked his head, looking at the Chief made him wonder what deputy Barney Fife might have looked like with a promotion and a couple of hundred pounds. "Well, that's a two-edged sword, Chief..."

"Sure enough, DeWald." The chief resumed his gum chewing. "But there are probably folks out there who have seen these boys stabbing these whales but they haven't come forward because they don't consider it to be a big deal. Now, if those same folks were to know these whale killers were killing innocent women, well...well, they might come forward. You hear what I'm saying?"

"Of course. But bear in mind, these guys are probably going tens of miles off shore in the middle of the night. The odds are nobody is gonna see them."

Laykin shifted in his chair. "We don't know they—"

"What's this *they* stuff?" the Mayor asked.

Laykin answered. "Mayor, killing a whale is not a one person job."

The Mayor nodded. "I guess I hadn't thought about that."

"Anyway," Laykin continued, "we don't know if they're killing these whales in the middle of the night, but if they are, so what? It's not a matter of how difficult it might be to catch them in the act, it's a matter of whether somebody might have witnessed the acts, regardless of the odds."

The Chief leaned forward towards DeWald. "They had to tow the things back, right? Wouldn't that look a little odd if someone were to witness it? I mean, if it were you, wouldn't you do a double take?" The Chief faced the Mayor. "Besides, by informing the public, it might get our letter writing folks off our backs."

DeWald nodded slowly and brought his hand up to his chin and began caressing it. What Laykin and the Chief had to say made a lot of sense. But he knew they had to be careful in revealing what they knew—

37

once the information was out, it was out, and the leverage it apportioned them would be completely spent. So how could they have their cake and eat it too? Of course…"There is a compromise here…"

"We're listening," the Mayor said.

"We could put out a reward for the identification of the whale killers. That way we may get a witness to come forward without having to disclose all the facts in this case, especially the link to the murders—"

"I must be missing something," Laykin said jumping in. "What is so damn awful about letting the public know about the connection?"

DeWald rubbed his forehead. "It's a key piece of information that we need to keep in our cards, we must have some mechanism for gleaning out suspects."

"We've got other information that we can use for that."

"True, but you should never lead with your trump card."

"For heaven's sake, we're not playing bridge—"

The Mayor spoke up. "Reward…We will probably get hundreds of people keeping their eyes out for our whalers."

Laykin chuckled. "There's only one problem with this fantastic idea. Where do we get the money?"

"How about a bake sale?" Thomas injected snidely. "Sweet."

"Agent DeWald's idea has merit, but the money is an issue," the Mayor said.

"This *is* New Bedford, right?" DeWald asked.

"No, it's San Francisco!" Laykin answered. Thomas and the Chief chuckled.

"What better way to show New Bedford is with the times? You can show the world that although you hold your whaling heritage dearly, you understand that times have changed, that things are different now, that you are with mainstream society. Take the bull by the horns. Express your outrage to the world. Have your citizens make donations to a fund! You'll get national press over this."

The Mayor and Chief both wrinkled their foreheads. "It could work, I guess," the Mayor said after a few moments.

Laykin turned red. "This is insane! We're not here to get national attention, we're here to solve murders!"

"Settle down, Laykin," the Chief stated calmly. "We can do both, what's the harm in that?"

Thomas leaned back in his chair and placed his hands behind his head. Laykin shook his head and said, "What about the poor woman out there that, if she had been informed, might have taken precautions and avoided becoming the next victim. Do any of us want that on our conscience?"

The room fell silent. DeWald got up from his chair and moved to the window. "I suggest," he began, "we inform the public that we believe the strangulations are related, and more than likely the work of a serial killer, but provide no further details. The first two victims were loners and we kept their stories relatively private, but the latest one is widely known, we can't hide this one, and Laykin is right, nor should we."

"What do you think, Laykin?" the Chief asked.

Laykin shook his head and shrugged and muttered, "Whatever."

"Good. We have a plan then," the Mayor said. "I'll get the City Manager on the money raising, and Chief, why don't you call a press conference to let the public know about the serial nature of the strangulations." The Mayor got up from her chair; an indication the meeting was adjourned.

DeWald grabbed Laykin and Thomas. "Let's talk shop shall we."

The three men met back at the Police station behind closed doors. The contents from the case folder were dispersed across a yellowed wooden table.

"Has the last victim been positively identified?" DeWald asked.

"Rueger, Kathy," Thomas said.

"She was a local newscaster, correct?"

"She was that," Laykin answered, his tone calmer. "And an investigative reporter. Apparently she was damn good at it too. Uncovered dirt just about every week on her show—"

"So lots of enemies..." DeWald said, thinking aloud. "Let's get tapes of her most recent shows."

"Why the most recent?" Thomas asked.

"Odds are her killer was someone she recently upset."

"If the killer had any brains at all, don't you think he'd wait until things cooled off before trying something?" Laykin asked.

"Assuming one of her interviews was a trigger, and it might not have been, then our killer would have reacted to something she said recently, I don't think he's the type of person that would wait."

"Junior," Laykin barked, "go down to the TV station and get the tapes." Thomas nodded and left the room. "He's a good kid you know."

"Attention deficit disorder, I'd say."

"Just eager."

Falling into the chair stationed across from Collin's desk, Bennett looked around the office waiting for Collin to finish what he was doing on his desktop computer. In the corner near the window, next to a plant dried through neglect—crisped by the sun while placed on the inner window ledge—stood a floor to ceiling bookshelf, skinned in brown veneer that bubbled in patches from spilled liquids, the only bookshelf in the entire office close to having its shelves filled with books. In it he had technical books ranging from software to hardware, vendor books used for pricing, and his literature: dank paperback classics with bent covers, marred by edge-wear, spread across three shelves starting with Shakespeare and ending with Dostoevsky.

In the middle of the office, was his desk coated in a blizzard of papers, save for a clear plastic gray outbox, that stood erect over the mess, yet it too was covered in an undisturbed film of dust. A chocolate in the shape of a vendor's logo and covered in silver foil, rested beneath a memo on the desk just in front of Bennett. It stood alone as a discrete recognizable object. Bennett removed the memo and picked it up and flipped it over. The date printed on the foil was a few years back. "Are you ever gonna eat this thing?"

"It's a decoration thing, not to eat," Collin replied, not removing his eyes from the display on his computer.

"How much longer you gonna be?"

Collin rolled his eyes. "Alright, Bennett, how can I help you?"

"I need a name. Someone good at encryption."

"Why so?"

Bennett placed the candy back on Collin's desk. Collin eyed it and picked it up.

"That *whale* file we found the other night—the data inside is encrypted."

"So?" Collin peeled the foil back from the candy and scrutinized the contents, rotating it like a chicken on a rotisserie.

"I want to know what this program does, but I need help with the encryption."

Collin bit a chunk out of the circle at the edge of the logo, chewed a little, pressed his lips together and nodded. "Not bad. Good vittles."

"Should I call the hospital now or later?"

"Well, as I understand it, Chocolate never goes bad."

"The encryption person, remember?"

Collin set down the candy. "Sandra Caulfield, she's good."

"Where do I find her?"

Collin made a few clicks with his mouse and read off a phone number. "Sure it's worth the effort?"

"Maybe, maybe not, but you know I'm always up to a challenge."

Collin smiled. "Well, that I know Bennett, that I know."

As Bennett walked back to his office with Sandra's phone number, he recalled the story he had once told Collin. He was a sophomore at the

University of Massachusetts. Dr. Gruber had made the pronouncement, as he had for the last twenty years in his Programming Algorithms course—after covering code tuning and optimization—that any student capable of producing code that executed faster than his for the *mouse in the maze* problem, was excused for the remainder of the course and would receive an *A*. He allowed the student to select the platform; after all, portability was important. No student had ever risen to his challenge. Bennett felt an immediate surge of energy; his roommate, Greg, said forget it.

"It's fixed…you can't win," Greg had said. That night, Bennett sat still, eyeing his modem, watching the send and receive lights blink like a marquee in Las Vegas. Hours later, as Greg watched from his bed, eyes half-open; the lights ceased to blink and Bennett thrust his hands in the air.

"Jeez, what are you doing now?" Greg asked, half-asleep.

"I have the answer."

"And it is…?"

Bennett refused to tell, but a week later the challenge took place at the front of the classroom. Dr. Druber placed a VT100 terminal on a desk and faced the screen to the students, whom closed themselves around. Dr. Druber completed the compilation of his code and removed the source—a master copy was in a secured locked safe. Bennett did the same with his code. Dr. Druber executed his code and received a time close to 23 seconds—a great amount of time, but a very complex maze. Bennett ran his and received a time of 21 seconds. The class went into an uproar. Greg hoisted Bennett to his shoulders and paraded him around the room.

It was years later when Bennett revealed his secret to Collin.

"So what did you do?" Collin asked.

Bennett blushed. "I altered the compiler."

"You what! You're bulling me." Collin's mouth dropped.

"When it compiled his code it found his name and introduced about five seconds of latency into his program."

"How did you know his name would be there and how come he didn't notice degradation from years past?"

"Standard format for code is to put your name at the top of the file, something Dr. Druber was obligated to follow, and he couldn't be sure about the degradation because the operating system had recently been upgraded. He thought this was the cause. He even contacted the vendor—can you believe it?"

Bennett laughed in silence at the memories as he walked into his office. In retrospect he wished he'd beaten Druber fair and square, and perhaps he might have if it hadn't been for the incident.

He placed the paper with Sandra's phone number on his desk and dialed the number. Sandra answered after a couple of rings and Bennett introduced himself.

"Any friend of Collin is *not* a friend of mine," she said.

Bennett stumbled for a reply.

"—I'm kidding," Sandra said laughing out before Bennett could recover. Her laugh sounded like a laugh he recalled his third grade teacher having—a delicate, light laugh, bound to dither to silence. Third grade—from what Bennett could recall—was a good grade; he felt comfortable with Sandra.

"Well, that's good news," Bennett said, laughing with Sandra.

"How can I help you, Bennett?"

Bennett pulled the printout he made that morning of the source code for the *whale* file. He went on to explain in detail what he had done so far in his attempts to unravel the mystery.

"Interesting," Sandra said after a few seconds. "Somebody has gone through a lot of trouble to make it as difficult as possible for someone else to figure-out what the program is supposed to do. Most folks probably would have been deterred by now, either by a lack of the required knowledge, or lack of interest in pursuing something locked up so tightly. As a matter of fact, I think they would have deterred about 98 percent of the people that would have come across this thing. But it sounds like they may have met their match now."

"Glad to hear you're very confident."

Sandra giggled. "I was talking about you."

"*Me?*"

"Yes, you must be very driven."

"You've figured me out already, I see," Bennett kidded.

"Why don't you send me the encrypted portion and I'll analyze it for any patterns I might recognize. I'm sure you realize I won't be able to tell you much if anything at all."

Bennett nodded as if she could see him. "I know Sandra. If nothing else, can we brute force it?"

"Have you got a Cray in your garage? Or how about a hundred or so PCs we could hookup to work in parallel?"

"Will a laptop do?"

Sandra laughed. "You realize that even if you decrypt the data you still probably won't know what it is—could be an image or just plain binary data."

Bennett realized Sandra did not have much of a background in coding. "That really won't matter. The program will run and I'll see the results—that is if there are any."

"So are you telling me the program knows how to decrypt the data?"

"Some parts. But the payload data, I don't think it has a clue how to deal with."

"Hmmm, this is very interesting. You've definitely pricked my interest. E-mail me the payload data."

Bennett copied down Sandra's e-mail address. They agreed to talk in a couple of days and share what they had learned. Bennett thought about the great amount of work that lay ahead. The decompiler had misinterpreted some instruction sets and been confused by others, perhaps lost in the quagmire of code compliance amongst the many standards available. And the debugger, its output a tedious stream of arcane codes, revealed operations performed by the program at glacial rates, causing Bennett to summarily decide that most operations were ancillary to the delivery of the encrypted payload. He wondered if it was worth it. He had no explanation as to why he found himself taking the situation so personally. He thought about investigating his motives and trying to get to the bottom of his obsession, but then thought better of it—that part to him was not worth it, the other, the enigma, well, probably so. It occurred to him he had gone an entire ten minutes and not once had he thought of Tanya. That of course changed, the moment that he did, and the detour he had taken returned to Tanya Street.

Patrick removed his back pin from the cribbage board and slid it beyond his front pin, 24 holes forward. "Dead hole."

He sat across from a crusty gentleman with white stubble coated cheeks leaning back in a chair. The pub was not open yet; it was still early. Known to a few as the *Bluefish*, the pub appeared as a pet entry to a towering row of three-story abutted multifamily housing units, the opaque black picture window with a tiny pink pig in the center was all that identified the pub. Inside, a soda fountain converted to a bar, and a few metal tables with Formica tops loosely coordinated with metal cushioned chairs, only three chairs matching, scattered on a beige floor of chipped tiles. The current owners purchased it from a man who tried to make it in the barbecue business. Not much was done to change the place when it exchanged hands, other than the sauce was replaced with the *sauce*. The mold laden smoky scent of hickory still lingered, this despite many scrubbings by Patrick, a co-owner.

"You're nothing but lucky," Karl Patterson, Patrick's business partner, contested.

"Beating you seven times in a row is not luck." Patrick shuffled the deck of cards. "Just call me the kid. Many long hours of playing, that's what it takes."

Karl chuckled. "Is that why we play everyday an hour before opening?" Karl pulled a cigarette from a pack on the table. "How's that wife of yours?"

"Candice?"

"No, Farah Fawcett."

Patrick smiled wryly. "Nagging, as usual. Says I should sell my part of the bar to you and get a real job. Yeah, right. Says I'm not making enough money.

"Is that right?"

"That's right." Patrick shook his head as Karl lit-up. "Look at your skin, it looks like freaking leather, man. You know why?"

Karl placed his lighter on the table. "Of course. It's because I smoke. But I love to smoke. It's the only pleasure I've left in life." Karl took a long drag and got up from his chair. "Do you know the only reason I converted this place into a pub was so I could smoke in here? Hell, if I'd left this a restaurant, I couldn't smoke!"

"Sit down Karl, I haven't won yet. And the decision to make this a pub was ours, not yours."

"Hell, I give the game to you."

"What a wuss, you can still win."

"Yeah right, just like I did the lottery last week."

"Just finish the game, Karl," Patrick said, pointing to the board.

"You're so anal about this—"

"Am good, aren't I?"

"Yes. But so what? It's not like you're going to the freakin' cribbage world series or something."

"Play…"

"Whose crib?"

"Who dealt?"

"You—"

"Then whose crib is it?"

"Alright, you don't have to be so damned sarcastic." Karl threw two cards on Patrick's side of the board. He blew rings of smoke. "You know your Uncle could kick your butt at this game."

Patrick threw his two cards on top of the two Karl threw. He could still hear in the background the bug-zapper just outside the screen on his Uncle Jack's summer cottage porch, as he and his Uncle played cribbage, seated in wooden chairs with a card table between them. He recalled how he longed for those evenings, or the Saturdays when Uncle Jack took he and

Bennett fishing or camping or one of any other activities so alluring to young men. Patrick looked up to the athletic man in the ball cap as his hero; the man who paid attention to him, the man who cared, the man who played a role that should have been played by the man who claimed to love his mother, but moved to the west coast, without a forwarding address to show it. But Uncle Jack left too, only not voluntarily, and some aided him…

Karl led with a card. "How long ago did he die?"

Patrick made fifteen and won the game. He removed the pegs from the board and slid them into the holes at the end of the board. He went behind the bar, grabbed a longneck, and leaned on top of the Formica counter. "Ten years ago."

"It was an accident right?"

"Some would say that…" Patrick said, popping his lips.

"You still blame Bennett?"

Volume dropped noticeably in the amber bottle as Patrick took a long swig. "Damn straight. If he hadn't been so stupid and gone out in the dinghy in that rough weather, Uncle Jack would never have been in the water in the first place. When Uncle Jack went out to save him, he found only the capsized dinghy, and our hero, Bennett, is lying across the seats in the upside-down dinghy, while Uncle Jack is thinking he's underwater. Uncle Jack dives in, and stays down as long as he can—meanwhile both boats drift away in the powerful current. It's not hard to figure out the rest of the story."

Karl looked in the mirror across from the bar seeing Patrick's blank stare.

"I don't think you can really blame Bennett for that."

"The idiot never should have gone out in that weather. Uncle Jack was the only father I had…"

"These things happen—"

The phone rang. Patrick left the bar, saying, "and someone is responsible for them," and went to the back office, shutting the door behind him.

"Turn to channel eight, man!" Paul screamed into his ear.

"We don't have a TV here, Paul. Damn you're here everyday, you know that."

"There's a reward out on us!" Paul's breathing was rushed.

"What are you talking about?"

"The city has put up money as a reward for finding us!"

"Why?"

"For harpooning those damn whales, man!"

"Are you serious?"

"As a heart attack—"

"How much?"

"They haven't said—who cares anyway!"

"Calm down. If they had a clue who was doing this, they never would have had to offer a reward."

"Calm down! Hell, I think this is great. We're outlaws, man! Paul and Patrick—kind of has a Bonnie and Clyde ring to it, don't you think?"

Patrick rested the phone down by his side, Paul's tin voice still screaming. He sighed and brought the phone back to his ear. "Listen, Mr. sounds-like-I-just-one-the-lottery" he said, "we've got to handle this calmly—"

"I'm coming down; you and I are gonna party!"

Patrick heard a click and a dial tone. He felt a dull ache in the pit of his stomach; he knew that, at some point, Paul would succumb to his ego and brag about it. Lifting his hands to eye-level, they appeared white, the palms clammy, the fingers shaking. He combed his hair back from his forehead and felt dampness. He grabbed a key chain from his pocket and fumbled for a small silver key and when he located it he opened the draw. There appeared a glimmer, a reflection from the chrome barrel. He pulled the drawer out all the way, and collected in his hand Karl's 44.

Bennett yanked a hot entree from the microwave. The steam almost burned his skin as he pulled it towards him. He placed a hot pad under the plastic dinner tray and grabbed a can of soda. Using both hands, he carried the two into the living room, set them on the sofa table, and plopped himself on the sofa. He grabbed the remote and turned on the television when the phone rang.

"Kathy Rueger's dead," Collin blurted into Bennett's ear.

"What?" Bennett said, looking over at the television.

"They're reporting she was strangled, hold on—"

Bennett heard Collin's television through the phone. He had brought the remote with him so he started flipping channels until the voices on the televisions synced-up. A woman reporter, her hair flaying in the wind, was standing on a beach.

"...A husband and wife who were taking a walk here at Horseneck beach discovered her fully clothed body in the sand. We have attempted to reach the couple for comment but we have been advised they are prevented from discussing the case because of the ongoing criminal investigation. Preliminary reports are that she was strangled."

"Karen," the voice said as the television screen split and a woman at the news desk appeared in a window opposite the reporter at the beach, "do the police have any suspects?"

"She had a lot of enemies, Julie. As we all know, she was a very good investigative reporter and she discovered a lot of dirt in this town. But to better address the question, I have here with me the lead investigator on the case, Detective Laykin with the New Bedford PD." The camera operating the window with the reporter zoomed out and Detective Laykin appeared in the frame. "Detective Laykin, you heard Julie's question, are there any suspects at this time?"

The wind blew the hat from Detective Laykin's head, causing his gray hairs to spiral in the wind. He looked back to see it rolling down the beach. "Jim will get it," the reporter said, and a man was seen running in the background after the hat.

Collin's voice echoed through the phone, "jeez she really ticked me off, but I wouldn't want this to happen to anybody."

Bennett remained glued to the television, but managed to say, "I hear you—"

"Shhh..." Collin said.

"Your answer Detective Laykin?"

Laykin recovered from losing the hat, attempting to flatten his hair with his hand. "It's too early in the investigation to have established any suspects at this time. However, we are giving the investigation a top priority—"

"The Chief in a press conference today announced that Kathy Rueger is actually a third victim for this strangler—"

"What he said was alleged third. We have strong evidence that the three are related, but cannot definitively say they are."

"Why wasn't the public notified about the apparent relationship of the first two murders?"

Laykin pursed his lips knowingly, but quickly retracted the expression. "Unfortunately Karen, sometimes in order to solve a case it requires that we not release information to the public to maintain the integrity of the case—"

"So why are you going public—"

"A'int this bizarre!" Collin shouted.

"I wonder what kind of freak we have walking our streets?" Bennett asked.

"Deranged and motivated, I'd say."

Bennett thought for a moment. "You think they'll think one of us did it?"

"What? Why?"

"She made fools of us that's why."

"Us and at least half of New Bedford and she's just one of three, remember? We have no motive for the others. Hold on—"

Bennett heard Collin put down his phone and the sounds from the television go silent. He returned to the phone. "Talk to Elander tomorrow. I think he has some information you might find interesting."

Bennett wrinkled his brow. "About what?"

"The *whale* file thing."

"What specifically?"

"Sorry, Bubba, don't recall exactly."

Once Bennett got off the phone with Collin he returned to his dinner, now tepid. The news of Kathy Rueger's death made him feel strange. He felt a tinge of guilt for being so upset with the woman, but it lapsed when his mind entertained what Collin had said about Elander—what could Elander possibly know? He didn't exactly like Elander: a young and cocky kid who had more body piercing then sweat glands, and who had grown up absorbed in computers, as had most his peers, in touch with cyberspace from the womb, born with mice clutched in their tiny pink mitts. In cyberspace Elander felt comfortable—at home—while Bennett kept only a foot in the door, still requiring a certain sense of tangibility.

"Be careful what you say, Laykin," DeWald said aloud to himself as he rested on one of the double beds in his hotel room, watching the news reporter interviewing Laykin on television. Flattened on the bed next to him rested a mangled bag from a fast food burger place. He eyed the television fearful of every word barked from Laykin's mouth. He had personally been involved in a case before where talking to the press proved a phenomenal mistake. He sighed with relief once the story concluded. Laykin had done a pretty good job handling it, as a professional should, with careful consideration of the touchier aspects of the case.

The bedsprings squeaked as he got up from the bed, he drudged along in near stupor, walking over to the round table by the window. The droll, stale, Damn-I-hate-this-place, atmosphere of the hotel room shorted-out what motivation he might have had from within the very synapses of his brain. The room was lined in striped wallpaper that shifted like a highway after an earthquake, the ceiling coated in bland stucco—all of which he had almost memorized including that part of the ceiling over his bed that eerily resembled a silhouette of Abraham Lincoln—the brown neutral colored carpet, the ever-present sterile smell from the daily cleanings done by a short Puerto Rican woman he called Ricky—his senses were lethargic, if not completely numb. He leaned over the table and grabbed a VCR he rented earlier in the day. Next to it on the table were the tapes from *Kathy's Korner*.

As he worked his way through the hookup of the VCR to the Hotel television and its rude security-related gadgetry, his wife, Linda, popped in his mind. He felt bad for missing last weekend's return trip to their home in Boston. He missed her. Chaps, their yellow Labrador, was keeping her company and standing guard, but he should be there too, after all, what kind of a relationship could they have if he was never there? He realized he had to start making progress on the case so that he could finally tell her that the end was in sight.

It took him over fifteen minutes, but he figured out how he could hook up the VCR to the Hotel's television. He looked over at the table and the stack of twenty-five tapes, and shook his head. Each tape covered three episodes. Well, it was his idea, and Laykin and Thomas were presumably viewing their share. It'll take me all night, he thought. DeWald overcame his begrudge, and slid the first tape in and lay back on the bed, propping his head with three pillows.

"...Good evening. I'm Kathy Rueger and this is Kathy's Korner. Tonight my guest is Shirley..."

"...Tonight my guest is Paul..."

The optic range of DeWald's eyes narrowed and later closed. Chaps was running in the backyard after a soccer ball as Linda and he swung on the back porch swing, the sunset enhancing the glow of her smile.

"...Tonight my guest is John Phillips, president of..."

Chapter IV

A prickly sting banded the back of Bennett's neck as he secured the top button on his shirt. Who would have thought the sun, at the latter hours of the day, a mere soft benevolent orange glow, could inflict such a scorching? A jaunt past the mirror in his entryway, the early evening before, made it clear that his skin was far too pallid to appeal to any woman, let alone a woman of Tanya's caliber. He grabbed the chaise lounge in the backyard and moved it to a section that still carried direct sunlight and lay sprawled for hours, rotating from front to back and vice versa, before falling asleep on his stomach.

Now he felt the agony. To offset it, the morning made for a promising day. No clouds and a light wind from the south—the type of spring day he recalled with fondness from his childhood—the kind where one walked under the blossoms of pear and cherry trees, where daffodils lighted a walkway, when a child can almost see the last day of school on the horizon. The weather, and lunch with Tanya, combined to lift Bennett's spirits. He threw his briefcase in the trunk of his car and set out for Pequod.

New Bedford, not long from its historical roots, was an interesting drive for him. Chestnut and magnolia trees lined the roads much as they had for the Quakers a century earlier. It was Dad who said so one day, when in route to the hardware store—a young Bennett in the passenger seat—to pick up balsa wood for the model ship Dad was to build. An ambulance screamed by in the present, heading in the opposite direction. Ignoring the piercing sound, he could see his dad, his arm resting in the opened car window, leaving every once in a while to adjust his dark Ray-Ban sunglasses.

Son, he said at the time, if you had taken this route in the heydays of New Bedford whaling you would have passed City Hall. It had a motto of lucem diffundo. Can you pronounce that? It means we diffuse light. You

50

know, from Sperm whale candles and lamps, and such. Just beyond the City Hall you had the Parker House, Bartlett's Market, Lowden's Millinery and Liberty Hall.

Dad took great pride in New Bedford. Knew everybody. A drive downtown produced no less than ten "good mornings" to so-and-so from the car window to folks working out in their yards or chatting outside the Post Office or walking their dog. Waldo, planted daily in a rusted card table chair outside the Sandwich Pharmacy, usually with the Times between his outstretched hands curtaining his face, would mumble a reply back to Dad without so much as shifting the paper below eye level to see who might have greeted him. He just knew the voice. Everybody knew Dad.

Turning left at 7th on to Spring Street, Bennett looked to his right, as he did every workday morning, at the black window with the pink pig. Both Patrick and Paul's cars were parked out front. Too early for them, they must have passed out and spent the night there. Bennett hung a right on 8th and after a few blocks pulled into Pequod.

In the kitchen area Collin burst out laughing when he spotted Bennett. "Did you attend the Kathy Rueger School of sunbathing?" he said, close to hysterics.

Bennett wrinkled his brow.

"What, you don't get it? They found her on the beach—get it? She was there for hours—toast, right?

"Very funny." He swore he could see his pink face in the reflection of the coffee he poured in his mug. "Just adding a little color, that's all."

Will and Adam were soon on the scene.

"Jeez, you're looking like a freakin' boiled lobster," Adam said.

"It's not that bad," Bennett said as he studied his outstretched rubicund arms, "it doesn't even really hurt." He thumped his arm and winced.

"Yeah, right," Will said.

"Y'all, looks like a Tanya tan to me," Collin said.

Will furrowed his brow. "Who's Tanya?"

"It doesn't matter," Bennett said. "I need to see Elander, anybody seen him?"

Adam said Elander was probably in his office. Bennett was hoping this wasn't the case—the room was uncomfortable to him. He left the kitchen area, the jokes still flying behind him—something about neon—and stopped at office 1503, where a year earlier Elander had worked over a weekend to replace the wooden door with strings of beads. The beads clicked as he pulled them enough apart to stick his head through. On the floor he found Elander, Alyson, and Peter, encircling the rug in front of Elander's desk, holding hands, sitting in yoga-like positions. All eyes were closed as a cinnamon smelling incense burned by the shaded window and wafted throughout their meditation zone. Three strategically placed lava lamps—providing the only illumination in the room—each capturing a different primary hue, twirled their innards from the desk, bookshelf, and at the window by the incense. Bennett noted that all three meditating wore bell-bottom pants. Elander wore his standard attire: a Nehru jacket. Alyson, an African-American, had a large Afro. Elander looked-up to see Bennett's darkened face peering down on him.

"How may we help you?" he said, closing his eyes and returning to his meditation.

Bennett wondered why he put up with it. Yet all he had to do was observe the room. Elander's walls were coated in certificates and awards. Alyson and Peter were not far behind. He always leaned towards the standard answer: my generation was odd at its time, so it's okay for this generation to be odd. The crazy thing about it was they were mimicking his parent's generation. Like the lava in the lamp, it must go in cycles.

"I can come back later," Bennett offered as a courtesy.

"No, now is okay. We were just ending our closing meditation."

The three unlocked their mutual grip and stood. Bennett stepped back while Alyson and Peter left. Elander went to his chair. Bennett knew to get a foldout chair from the corner of the room near the bookcase. With caution, he teetered on the canvas seat supported by three spindly wooden legs crossing themselves halfway down to the floor.

"How can I help?" Elander repeated. The silver rings in his nose, ears, and eyelids, cast a soft multi-colored glimmer from the lamps. He looked like a Christmas tree when the lights in the room are extinguished and all that remains are the ornaments, tinsel, and colored lights.

Bennett thinking: what a freak. "Collin tells me you've got some information concerning the file we found on the system the other night."

Elander grabbed a Rubik's cube from his desk and began its manipulation. "You referring to the *whale* file?"

Bennett shifted his weight in the chair and almost toppled. He regained his balance by gripping Elander's desk. "Yes, what else do you know?"

Elander made his last adjustment and the cube was complete. "Why are you pursuing this?"

"It's a challenge—someone's testing us, can't you see that?" Bennett eyed the cube. "How is it so different from your cube?"

"Unraveling the cube is a positive exercise, an affirmation of knowledge. Competing with others is a negative exercise, a defamation of knowledge. We should turn the other cheek."

"And what if this thing takes down our entire network and we're out of a job?"

Elander shrugged. "So we get a job somewhere else..."

The edge of impatience crept into Bennett's voice. "Look, someday we'll debate philosophies, but not today. What do you know?"

Elander shrugged. "In my travels the other day on *aider route*, I came across a gentleman—I think he traveled from San Francisco—who asked if I had encountered, or chatted with anyone, concerning the discovery of a *whale* file on one's network."

"And your answer was..."

"At the time I had no knowledge of our encounter, so of course I said no."

Bennett hated to ask, but knew he had no choice. "What is the plebian definition of *aider route* and can I still find this gentleman there?"

Elander rolled his eyes. "It's a newsgroup, and yes, the gentleman— Griggs, I think was the name—left a posting detailing his encounter."

Bennett grinned. A newsgroup was nothing more than an electronic bulletin board and somehow it had acquired a fancy French name and was a place where one could actually encounter other individuals. Cyberspace was what it was, global electronic communication and the only things living there were programs, some dumb, others slightly smarter, and some might even be considered clever, but none human, and certainly not magical.

Bennett thanked Elander and folded the chair back up, noting, with his eyes now properly adjusted, that the canvas seat sported a yellow smiley face.

"If you hear anything out there, let me know," Bennett asked, pulling the beads apart to leave the room.

"Leave it alone..." Elander whispered under his breath.

Static filled the television screen when DeWald set down his cup of coffee and moved over to the VCR to hit rewind.

"So what'd you think?" he asked, looking back at Laykin, who was working on a second donut, and at Thomas, who was eyeing his cigarettes.

Laykin wiped Boston crème from the corner of his mouth. "I don't see the connection between these bizarre murders and some guy whose running his liquor store illegally."

"The guy was really pissed-off," DeWald replied, holding his hands out.

Thomas sniffed. "I know the guy—I went to school with his son. He's a moron alright, but he wouldn't knock off somebody—"

"Look," Laykin said, flustered, "whoever killed Rueger, killed the other two. Maybe this guy has motive for knocking-off Rueger, but what motive could he possibly have for the other two victims. I think we have to assume the three murders were all committed by the same killer."

"Killers have been known to break their patterns under extenuating circumstances," DeWald said.

"What's that supposed to mean?"

"The Rueger murder could have been a rage killing; the first two murders might have been perpetrated under a different set of motives. Thus we could have one killer and two different sets of motives."

"Could be a copy-cat job," Thomas said.

DeWald smiled. "Thank-you, Junior."

"For what?"

"For providing an excellent example as to why we are not disclosing all the details of this case to the public. Without the information we have, how can someone possibly copy the crime?"

"It could be done." Thomas tapped out a cigarette from its pack and popped it in his mouth. Than he pulled it out. "The copy-cat killer might know the real killer—kind of have inside info…"

Even Laykin rolled his eyes.

DeWald placed the tape back in its container. "From what I saw, not a guest on her show stands out as a more likely candidate than the others when you throw the other two women in the mix."

"Did you watch her last one?" Laykin asked.

"Was it on one of my tapes?"

"Yep."

"I suppose I watched it then."

Laykin took the remainder of his donut in his mouth and spoke while still chewing. "I think you missed it. I happened to catch it the night it was shown. The guest is from an organization that will most likely yield our suspect."

"Who was the guest?"

"John Phillips."

DeWald returned to the table and quickly flipped through his yellow legal pad. The name did not appear in his notes. He wondered how he could have missed it, and grimaced when he recalled he had dozed for a few minutes. He shook his head. "Somehow I missed the guy."

"Doesn't matter, us locals are on top of it. We've already set some things in motion."

"Oh." DeWald's eyes widened. "How so?"

"We've got Detective Allbright working undercover."

"In what capacity?"

"John Phillips belongs to the New Bedford Historical Whaling Society. This suspect, or anyone of the other members, ties together Rueger with the whaling angle, which then ties in the other women. Detective Allbright will be keeping an eye on him and other key members." Laykin pointed to DeWald's stack of tapes. "You might want review the episode."

Thomas smirked. "Sweet. Smoke break gentlemen?"

DeWald colored. In his ten years as an agent he had never been so careless. If Laykin hadn't watched the last episode a potentially major break in the case might have been missed. He wanted to kick himself or shake himself and drive home the point of what a major mistake he had made. He promised himself, no matter what, it would not happen again.

He paused from his walking to lean over and read the headline of a local New Bedford rag placed behind the glass of a newspaper dispenser. *Boston Strangler Returns?* Bennett started to scan the first few lines of the article, was annoyed at a hint of sensationalism and so stopped reading, and continued his walk to Macs where he would join Tanya for lunch. When he arrived at Mechanics Lane, he frowned at the window display for the *Shirt Off My Back* T-shirt shop. The phrase *"The strangler was here"*, was printed across a resplendent tie-dyed T-shirt with a white collar that had a rope printed around it. A mania had started, and there was always money to be made on a mania. His eyes traveled to a female mannequin draped in a long T-shirt, his mind shifted to thoughts of Tanya.

He mulled over what he could remember from the night before. She was from Madison, Wisconsin and moved to New Bedford a month or so ago. He told her she didn't look like someone from Wisconsin and she wondered what someone from Wisconsin looked like. He remembered the giddiness and how it inebriated him such that he laughed at the slightest hint of humor in her words. He kept telling himself to sober up.

He was surprised to find out she was a buyer for the Millers grocery store chain. She had a big chunk of territory that spanned most of southeastern Massachusetts. She was unlike women he had known in the past, something he attributed to her charisma, and her out-of-band mannerisms that tickled his sensibilities.

Bennett arrived at Elm and turned south, halting at the storefront with the Americana picture window covered by a three-foot hamburger— sesame seed buns, lettuce, mustard, and ketchup (flowing like lava)— painted on the glass. Peering between leaves of lime green lettuce, he didn't see Tanya inside. He said a quick hello to Mac—who stood behind the vinyltop counter, wearing a thick froth of sweat on his forehead and a brown stained white apron that hung from his neck, and was holding a grease-dripping spatula—and grabbed a window booth. A large bag of lemon-yellow fries painted on the glass shaded one side of the booth. Bennett sat on the sunny side. How much worse could the burn get? He looked between the painted fries on the window and out to the street. His heart pounded as he watched people walking along the sidewalk shoot into view, not knowing when one would be Tanya.

"You want something to drink?"

Bennett jumped from his seat. A waitress, wearing a Mac's T-Shirt and white shorts, stood at the end of the booth table, smiling. Bennett waved her off and when he returned his stare to the street he noticed a flash of amber disappearing behind the fries. He turned and watched Tanya walk in the door.

"Sorry I'm late. I was making last minute arrangements to go to Nantucket," she said, as she slipped into the booth. The Freesia scent was back. "It's good to see you." She winked.

"Nantucket?" Bennett asked, open-eyed. A weekend excursion, he thought. His heart sank. She would be meeting somebody.

"I'm making my initial rounds to all of the locations in my area."

Bennett sighed. Maybe there still was a chance she was not seeing somebody. "Have you been there before?"

"Haven't had the time."

"You'll love it—cobblestone streets, old street lamps, very colonial. I'm surprised they allowed a Millers on the island."

"I'm sure they were pissy about it. Millers probably slapped a couple of columns out front to make it conform."

"They probably had to do more than that."

Tanya pursed her lips and broke-up in the midst of her thinking. "Sounds like something my dad could have pulled off. He's a glib talker—bigger than life." She looked into Bennett's eyes. "Any man I end up with has to be able to fill his shoes."

"Okay…" Bennett said sheepishly. "Your own litmus test."

Tanya leaned forward and squinted as she looked Bennett over. "You're burned!" she blurted out.

Bennett puffed a dismissive laugh.

"That's too bad, the sun is really bad for you," she continued, "I don't think people should get too much sun." She held her arms out and looked at them.

"I agree totally. I fell asleep outdoors. I was gonna only stay out a few minutes."

Tanya studied the menu for a moment and Bennett pretended to do the same. She asked Bennett for a suggestion, and a few moments later Bennett was ordering a Mac special for both of them. She glanced over at

Mac behind the counter, who was now leaning over the grill flipping patties. He swiped his apron across his forehead. Tanya winced. "Who's that guy?"

"That's the owner, Mac. He's a downtown institution. All the business types come here for lunch—have for decades."

"He sweats like a water sprinkler."

Bennett raised one of his brows. "It's what makes his burgers special, it's his Mac sauce, the secret ingredient."

Tanya's jaw dropped, and Bennett broke out in laughter. When he caught his breath he said, "I'm kidding…Macs been here for years, trust me, the food is fine."

Tanya frowned and shook her head. "I'm not so sure." She held her hands out and looked at her nails.

"I've eaten here forever, it's okay, really."

"Well I know some food inspectors and I've heard otherwise."

It was Bennett's turn to look surprised.

"Gotcha!" A smile returned to Tanya's face.

"Touché…"

Tanya rearranged herself in the booth. "So, this strangler thing is kind of creepy don't you think?" she asked and took a sip from her just served soda.

"Yeah, I guess it's becoming the talk of the town. Are you worried?"

Tanya nodded. "Yeah—being as a single woman and all—but hey, I'm tough." Tanya formed a muscle with her arm.

Bennett grinned. "So, you don't really have someone to watch out for you? A friend?"

"No, haven't really had time to meet many people here. My family is all back in Madison, so I'm pretty much alone."

"Anytime you need an escort or just want someone around, let me know."

Tanya giggled. "You're so sweet. Who'd have thought I'd meet a nice guy right off the bat." She winked. Freesia, again.

Bennett felt himself blushing. Thank goodness for the sunburn.

"I'm pretty independent though," she continued, "I work in a tough environment, vendors and all. If you let them, they'll eat you alive."

"Yeah, I can imagine—"

"And what about the harpooning of those whales? They're such magnificent creatures." She lifted her brows. "It's an abomination. You know what I'm saying."

"But trust me, stuff like this is very uncommon here in New Bedford. Before this stuff, the most unusual thing I read in our local Police Blotter was the woman who called the Police insisting her neighbor had a dead cat on the porch—it hadn't moved in days. When the Police checked it out they discovered the cat was a doorstop sculpture made of cement."

Tanya laughed. "Is that right. Now *that's* odd." She looked out the window and then at Bennett, the sun bathing his arms and legs. "I should be the one sitting in the sun," she said.

"No, no, that's alright," Bennett said, waving his hand. After a moment of silence, he dropped his voice. "I'm glad to see you recognize it's okay to be a member of our group. With everything going on and all, we might start to look like the bad guys."

"Your club is historical. I'm a big fan of history. This stuff has nothing in common with your group." She thought for a moment. "No one in your group would even consider harpooning whales, right? You know what I mean."

"No way."

"I mean, I guess every group attracts some weirdos. When I was in high school, a kid in our conservation club tried to burn down our school. So I guess yours might."

Bennett shook his head. "I don't think so. Besides, this town is rich with the whaling motif. There's Melville Street, the Seaman's Bethel, the Whaling Museum, and so on. It wouldn't be much of a stretch for anyone in the area to pick up on. Maybe it's a fad thing. Maybe some group, you know, bored by mere fishing, has upped the ante to make the sport more of a challenge. When you start speculating, you realize there are hundreds of possibilities."

"I guess you're right. They could even be boating in from Maine or New York."

The waitress arrived with the Mac specials. Tanya took a big bite and licked her finger. "Delicious," she said.

"Mac makes a tasty burger."

"I love the aroma. I especially like that hint of sweat. Yippee chi yea!"

All Bennett could smell was freesia.

"I'll leave you two be," Karl Patterson said, as he stepped outside the front door of the pub. "We need more peanuts."

Light flashed through the opening of the door and illuminated two men sitting at one of the tables. "The heat's on," Patrick said, covering his eyes from the light. "I figure we're only about three moves ahead right now."

Tapping his fingers on the tabletop, the man called Ahab said curtly, "So?"

"So? I don't want to go to freakin' jail, alright?"

The man called Ahab sat back in his chair and rubbed his chin. "For slaying the Leviathan—surely you jest—it's a God given right."

"I don't know where you live—"

"What's it gonna take, money-wise, to assuage your fears?"

"It ain't the money…" It was the money. And now Patrick knew was the perfect time to apply some leverage.

"The production can't stop yet. The journey we have embarked upon is far from complete. It's not unusual for a whaling vessel to remain at sea for years, navigating the South Seas from stem to stern until it's hull weighs full with whale oil and whalebone and sometimes an occasional nugget of ambergris. The life of the whaler is fraught with still seas, a boiling sun that blisters one's back, and endless hours of waiting, until the hark from the crows nest signaling 'There she blows!'—"

"I'm not a whaler from freakin' 1834, I'm just some drunk trying to make money with my boat, trying to keep my wife happy. Nothing more. But it's not worth a trip downtown."

The man parted his lips to a sliver and let loose with a hearty, but strained chuckle. His hand dipped generously into the peanut dish. Patrick looked close at the mystery man sitting across the table from him. He pondered the man's tragic face. The skin had a polished, but pockmarked, clay-like texture to it—at best guess the result of an incineration. The unblinking, yellowed eyes sculpted beneath the taut, pore-less forehead held shiny pools of ebony at their center. As he stared, the man grabbed a portion of the braided cord about his neck, pulling a silver ornament attached at the bottom from beneath his shirt. He carefully caressed the silver between his fingers.

"So what are you saying?" he said, putting the silver piece back under his shirt.

"I'm saying, let's cool it for awhile—"

"I'll pay you double…"

Patrick turned his head, slightly shaking it. He sloshed his lips from side to side and sighed. Inwardly he smiled. "One condition…"

"I'm listening…"

"I get to leave something behind."

The black pools twitched. A handful of peanuts shot through the slit between his lips. "Why?" Between crunches.

"Take the cops down a side path."

"Mislead them?" The slit of a mouth curved-up. "I like it. I knew I had the right man for the job."

Patrick smiled and nodded. "Just make sure you don't remove it."

"How will I know what it is?"

Patrick chuckled and looked around the room. "You'll know what it is, trust me."

The whites in the laundry basket were tossed into the washing machine. The heap might have remained untouched, if not for the fact that Bennett, when dressing that morning had reached for the ragged pair of underwear in the back of the drawer, the lone pair remaining from his high school days, his pair of last resort. As he dumped a heaping scoop of detergent in, the doorbell rang. He moved to the door and peered through the black wrought iron grid laced across a small window, near the top of the door, and observed an eclipsed orb. While questioning whether to open the door for a stranger, he heard a muffled voice.

"Bennett, it's Alfred, you home?"

Alfred? Who was Alfred? Bennett opened the door.

"Glad to find you home," Alfred Huffnagle said, extending his hand.

"Alfred...welcome," Bennett said blankly, shaking his hand, and then recalling the face from the last NBWHS meeting.

"Hope I'm not interrupting anything. Wife's got bridge tonight so I thought I'd take a walk." Alfred looked in the direction of the living room.

"Have a seat," Bennett said, motioning to a chair in the living room. "Just doing some laundry, no big deal." The two settled in the living room. He asked Alfred if he would like a drink but the request was ignored as Alfred ran his eyes across the room. Alfred's bony white legs dangled from his plaid shorts. His T-shirt dripped from his chest.

"A lot of whaling pictures. You must really be fascinated by whaling." He removed his eyeglasses and polished the lenses with a cloth he retrieved from his shirt pocket.

Bennett nodded. "Yeah, I find an unusual beauty in it. I think it adds a local historical flavor to the house." He looked over his shoulder at the pictures, the ones he had received from the home he grew up in. His mother didn't care for them, never really had, so she handed them over without hesitation. In fact, she had made it clear that she hoped someone would take them. He recalled how, after he had finished hanging them, his house felt so much more like a home and that somehow Dad was there, maybe not seated in the recliner, but in spirit.

Alfred returned his eyeglasses and continued his observance of the artwork.

Bennett bobbed his leg that was resting on top the other, and crossed his arms. "If you don't mind me asking, how did you know I lived here?"

"The member's directory—I picked one up last meeting. You guys had them stacked on a table at the meeting. I thought they were fair game."

"Of course. So we're neighbors?"

"I'm just about two blocks out. It's our summer place. Try to walk five or six every night."

"Miles?"

"No, blocks."

Bennett looked at the wall behind him. "Does the artwork bother you?"

Alfred waved his arm. "Are you kidding? I grew up with this stuff."

"Whaling?"

"Yes, I remember seeing some of those ships." He pointed at the wall.

"How old are you?"

Alfred grinned. "Old enough."

Bennett wrinkled his forehead. "You would have to be well over 100 years old." When he first saw Alfred at the NBHWS, he thought the man to be in his mid to late fifties. He figured the man went for early retirement.

"Well, you know, the ships—they were still around. Not the whaling. Hell, the Lagoda sits in the museum today."

"The Lagoda was never a real ship."

"Yes, or course. Probably have it confused with some other bark."

Alfred's eyes moved throughout the room and Bennett watched the odd man with curiosity. Bennett wasn't sure if the man was looking for additional conversation pieces or was just plain engaged by the setting.

"I talked to that Collin fellow a little bit," Alfred said, his eyes focused on computer magazines spread on a side table. "He must be from the South."

From the South? Bennett chuckled inside. He recalled the time at Pequod when a sister company in Huntsville, Alabama, yet another ISP, needed to borrow some expertise for the expansion of their lab. Collin was volunteered for a six-month stint. When he returned, he brought Southern jargon with him. "It's homey", he had said, "more like my nature. Don't think I was ever meant to be a Yankee. I'm a genuine Southern gentleman."

"No, not from the south. He took a liking to it after a visit down south—he's permanently adopted it."

"Can't say I blame him. There's something special about the South." Alfred rubbed his hands together. "Could I use your bathroom?"

Bennett nodded and pointed to a door down the hall in the direction of his bedroom.

"You know us old guys, can't hold it anymore."

Both chuckled. Alfred scurried off to the bathroom and did not reappear for over ten minutes. His hands dripping, he stopped in the hallway.

"There's a towel you can use in there," Bennett said.

"That's okay, air-dry, always the best way to go."

Bennett shrugged. "Okay, sounds as good as any."

Alfred eyed the hallway and looked at his watch. "Well I better go. Want to get home before the misses—she'll panic, if I'm not there. Care to join me on the remainder of my walk?"

The buzzer from the washing machine sounded. Bennett shook his head. "I'd better not. Lot's of laundry to go and I've got some work to take care of—"

"Exercise is good for the soul." Alfred's eyes traveled to Bennett's feet.

"You're right on that—"

"Got any rubber boots or surf casting equipment?"

"I've got some boots."

"I'd like to spin cast from the beach."

"You want to borrow the boots?"

He shook his head. "No, I need the equipment first."

Alfred turned towards the door and Bennett got up from his chair.

He let him out and waved as the thin man scooted down the drive, the sidewalk, and finally down the street.

After putting in another load, he clambered up the short staircase to the top floor. He moved into his office and turned on his laptop. He planned a quick journey to *aider route*. As he dialed in to Pequod he hoped the article would still be present in the newsgroup. It was his first, and only known clue, outside of what had been found at Peqoud. He searched on the word *whale* and immediately found his screen flooded with environmental group postings. He knew it would take an eternity to sort through all the articles so he began thinking of an alternate search. He tried to think of something that would involve the word *whale,* and include the context whereby the guy might have framed the article, but would preclude the environmental postings. He thought back to his initial reaction—that it might be a virus, maybe a Trojan horse. He tried another search, this time on *whale* and *virus*. He got back two postings—an environmental topic and the one he wanted.

Anybody out there run across a file named "whale"? It just appeared on my machine. It's a Unix executable but doesn't appear to do anything. I'm afraid it might be a virus (Trojan horse?), but it doesn't look like it did anything to my system. If anybody can give me more information on this I'd appreciate it. I've removed it from my system, but have saved it to a floppy. Thanks, Griggs.

Bennett clicked on Griggs' e-mail address to reply to Griggs only and not the newsgroup.

Griggs,

I've come across the very same mysterious file. I've done some preliminary analysis of the file but haven't got far. Whoever put the thing together did not want anybody getting in. I think somebody is playing the role of a rogue Unix guru and is challenging a select set of people to see if they can meet his challenge. My guess is you know your way around Unix and he has sent a challenge to you much as he did me. Let's team-up on him. I will attach a file containing the source code for his program (It's sketchy at best, I decompiled it) and my notes on what I know to date. Please reply ASAP with any of your findings.

Sincerely, Bennett

Bennett clicked on the send button and his e-mail went, its genetic bits beamed down the wire. Seconds later his phone rang.

"So when're you getting over here to help me setup my new PC—you know what I mean?" It was Tanya, feigning impatience.

DeWald laced his tennis shoes, sat up, and sighed. Did he really think he was up to running this evening? Next morning was out of the question; they would start early. The day had dragged long into the evening. Laykin brought by the lab report from the Rueger murder around noon. The two began combing through the data hoping a tickler would come their way—something to spawn some fresh ideas.

The sand crunched under his shoes—the sun having baked a crust on it during the day—as he stepped from the blacktop parking lot to the beach. The sun painted the horizon red, forming a backdrop for the islands, defining Woods Hole to Cuttyhunk. A few tall sails skating across the bay, cutting teeth into the horizon, completed the picture save for a herd of powerboats circulating no more than a hundred yards offshore. He paused to gaze at the boats, but paid them no significant attention. He seated himself on the sand with his legs outstretched. The sand felt warm beneath him. He leaned over each leg for a three count and then was back on his feet. Time to run. He positioned himself at the high-tide line and began his run parallel to the mild glassine like surf.

The physical evidence was disappointing. On the one hand it supported his earlier suppositions. The whales were harpooned at sea, more than likely aided in their floatation by barrels or something similar—a marine biologist informed him that humpbacks may or may not float once dead—towed to shore, affixed to the beach by very strong rope, and at some point in time, fed the victim. On the other hand the evidence added nothing new to the picture. Thomas returned with his findings on trying to identify the rope, the only tangible evidence left behind other than the victim. He reported the rope was a garden variety, a type sold in more than a dozen stores in the area. He had reviewed video from those stores that had cameras, and captured still photos of those making the purchases. He had interviewed the purchasers, at least those that could be identified, and found none of the group a viable suspect. As for the victims, all had the tattoo implanted on their forehead, and in all cases the ligature used as the murder weapon was determined to be thick fishing line.

The group of powerboats, now in the dark and defined by nothing more than a collage of green and red lights rhythmically undulating, formed a single line, and were headed in the direction of the shore ahead of DeWald. The scent of gasoline, a natural side effect of the outboards on the boats, wafted in off the water.

DeWald ran through the crime scene in his mind. His thoughts traveled from the shore, to the whale, to the rope, to Kathy Rueger. Her pallid stiff body wrapped in the body bag like leftovers in the refrigerator. Twenty-eight years of learning, experiences, memories—emptied, drained to oblivion, leaving in their departure a vacant vessel, without context, with no explanation—other than in living minds—for moments in time as small as the scar beneath her chin, for which he pictured her father coddling her and holding a cloth to it to stop the bleeding until they arrived at the doctor. Twenty-eight years squandered at the hands of some greedy beast, a dolt to all that is decent. Twenty-eight years snuffed out, inhaled by the murderer like some human narcotic, in order that his desires—reached only through extortion of life from a fresh kill—might be sated. All told, with the three, seventy-four years. And for what?

The murderer had made his last sacrifice, seventy-four was all he was ever going to get...

He allowed his mind to continue to wander the crime scene, stopping when it reached the white dental stone left by the forensic team at the base of the tree—the child's hand imprint—the one that kept appearing at every crime scene. Why was it there? What did it have to do with anything? Was a child somehow involved with the sickening crimes, or could it have been made with a doll's hand, as conjectured by the forensics people? What was its significance? It had to mean something to the killer, but what? When he had talked to his wife Linda the other night, her psychiatric opinion leaned in the direction of a lost innocence theme. This caused a flash thought of just how much he really missed her. But he was right back to the case and the more he tried to search for answers, the more fleeting the answers became.

The powerboats were anchored just off shore. On the beach, a fire threw flames and sparks into the night. DeWald could hear laughter and loud voices. He thought twice about turning around before he reached them and heading back to the car. But it occurred to him that his running was a tangible means of expressing his commitment to solving the case. He would not cut short his running, nor would he cut short his efforts on the case. He continued, running even faster. Someone from the fire shouted out and asked if he would consider joining them for a beer. DeWald declined as best he could grunt out between breaths.

Concentrate on the case. He thought about what Laykin had done, introducing an undercover cop into the NBHWS without consulting him at the time, and it upset him. The typical agent would have given Laykin a severe butt chewing, indeed, most agents were steeped in the prerogative, but it wasn't his style, the idea of Laykin's was a good one. It was expected Detective Allbright would be reporting any findings shortly—something, anything, to get a grip on perhaps a thread of evidence. He knew if he got a thread, he could unravel the case. The hardest part of any case was finding the place to start unraveling it.

DeWald turned at the dike and started on his way back. Against his better judgement he stopped at the fire to check things out. A ring of men were seated around the fire, a series of orange faces hanging suspended in the air, the dark obscuring their nondescript bodies. A collection of beer bottles, some empty, some not, lay in the sand about them.

"What are you fellows up to?" DeWald asked, in as much of an official voice as he could muster.

A bearded man squatted on a cooler spoke up. "It'd take a lot of money to get us out on a week night like this, but ten thousand dollars is a lot of money. We're gonna apprehend those whale killers."

"I see," DeWald said. Ten thousand dollars? New Bedford had come through with more than he had expected.

"You can join us if you'd like," said a man sitting Indian-style in the sand and wearing a bandanna.

DeWald shook his head. "What are you boys gonna do if you see something?"

The bearded man laughed and the others followed. "We're gonna haul them in, of course."

DeWald smiled. "These men could be very dangerous—"

"Got a gun in the boat. They give us any trouble we blow them away," the man with the bandanna said.

"How many of you have guns?"

The bearded man shrugged. "All of us."

"So lets see…you're drinking, you've all got weapons, and you're heading out to sea in powerboats to hunt down a potentially armed man or men—do you see something wrong with this picture? Don't you think there's a little potential here for a very bad situation?"

The bearded man shook his head. "No…"

"Listen boys, I'm an agent with the FBI." The men exchanged furtive glances. "I have no problem with what you're doing, but do me a favor, if you see something suspicious, radio the authorities, don't try to arrest these folks yourselves, or believe me, there's gonna be some serious trouble. Do all of you have radios in your boats?"

"I do," the man with the bandanna, said.

"Anybody else?"

The men shook their heads in unison.

"Great. Okay…" DeWald pointed to the man with the bandanna.

"Bill," the man responded.

"Okay, stick together, if anybody sees something, find Bill and let him radio the authorities."

The men in the circle nodded. Before DeWald entered his car he looked down the beach at the fire. Maybe this bounty thing wasn't such a good idea. He shook his head. God be with them tonight. If one doesn't die, it will be a miracle.

Not again, Bennett thought. He looked at the fuzzy red numbers on his alarm clock, taking a few moments to focus them in. Twelve-thirty. Not as bad as last time. He knew someone was messing with Pequod again or Adam never would have woken him. He picked up his radio and clicked the side button to talk.

"I copy Adam, what's up with the network?"

"Seagulls are swarming!"

Chapter V

Bennett navigated to Pequod down silent, deserted streets. Attempted penetrations against Pequod were normal, but usually limited to impotent port scans and known vulnerabilities from years back, long since addressed by security at Pequod. But this had changed. As he pulled in the lot he found Collin getting out of his car.

"Mo'nin. Ain't this getting old," Collin said, a cup of 7-Eleven coffee steaming in his hand.

"Somebody's got our number," Bennett said, pulling a baseball cap from his head, combing through his hair with his fingers, and replacing the cap.

Collin took a sip of coffee. "Think they're attempting a serious penetration, or just screwing with our minds?"

"I honestly think somebody is playing a game with us."

"Maybe we should stop playing, after all it takes two to tango."

"How does he know we've been playing so far?"

"The *whale* file never phoned home."

"We could have trashed it, that's not really playing."

"Maybe he considers that playing."

"I'm just gonna have to figure out what the purpose of the *whale* file is."

Collin mocked a chuckle and shook his head. "Why? Why are you so relentlessly pursuing this?"

Bennett looked surprised. "What do you mean? Why shouldn't I?"

"Face it, it was that head injury of yours. You have to keep proving to yourself that you're still darn good at what you do."

Bennett shook his head. "Did you ever stop and think that finding out what this thing does could be important? I mean, what is the deal here—you, Elander, Will—all of you just want to ignore what's going on—"

"We're not ignoring it, but you just might be blowing this thing out of proportion."

"Oh, so we wait until our entire network is blown away and then react?"

"Well, that's not what I'm saying. What I'm saying is before we go whole hog we closely monitor it until it does something more than sit there and look pretty, and if it does, we take action, and by that I mean whatever action is necessary to protect our network."

Bennett pulled his lips tight against his teeth and then released them. "That's fine for you guys, but it's just not in my nature to sit back and wait and see what happens."

Collin sighed. "Look, the odds are this thing isn't going to do anything. You know those gag handguns at the fair, the ones where you pull the trigger and out comes a flag with the word 'Bang!' printed on it. That's exactly what's going to happen here. We'll spend three months decrypting this stuff and executing it, just to see the word 'Gotcha!' flash across the screen."

"I'm doing this on my own time Collin. If that's the final result I eat crow, but I'm going to take the chance."

"Man, are you stubborn."

"It's who I am…"

Collin grinned and set his cup of coffee on the hood of his car. He extended his hand. "Truce."

Bennett nodded. "Truce." He watched his hat fly from his head when Collin flicked it with his index finger. "That is so lame."

"It's who I am. Well, before either of us does anything more, we need to checkout what we've got upstairs."

Bennett and Collin stepped into the brilliant white lab, both squinting as they entered. Adam looked up from a monitor, spotted them, and trotted over to greet them.

"Where's Will?" Adam asked.

"His wife is ill, not a big enough emergency to bring him in," Bennett replied.

"So what have we got?" Collin asked.

Adam studied his clipboard. "As I told Bennett on his trip in, these seagull files—about ten of them—appeared and then disappeared. Started out as one and that one spawned the other nine."

Bennett rubbed his eyes. "So what did they do?"

Adam smiled. "The good thing is we were watching the wire. It appears they sent encrypted traffic back out on the Internet through the mail port."

"Can't all be encrypted."

"Well not the headers, but the body of the messages were."

"Where to?" Collin asked.

"A server at a major ISP."

"The site has been compromised. I'm sure the messages were forwarded from there to the actual destination," Bennett said. "Do we have the traffic saved?"

"Sure do."

Collin cleared his throat. "So to summarize, a single seagull file ran, it created nine more, then all ten sent some sort of information back out to some unknown location on the Internet, and finally they were rendered asunder."

"...And, we have that information saved," Bennett added.

"But wait, there's more..." Adam said, sounding like a television commercial selling some inane product. He handed Bennett a floppy disk. "Compliments of *Wolf Spider*."

"Let me guess—"

"Yes, another *whale* file."

Bennett's jaw dropped. "I was gonna say *seagull* file."

"Got one of those too, at least an image of the running process."

Bennett took a chair. "This is getting stranger by the moment. It's like we've got seagulls swarming over a whale—"

"Seems our network is becoming Sea World," Collin said and laughed.

"Really, it's like somebody is trying to paint a picture, to replicate some sort of fantasy," Bennett said, leaning back in the chair.

Adam pressed a pen against his lips. "Maybe this really is a game. Maybe we've got some geeks out there using our network as a staging area for their games."

"Maybe," Bennett said, looking up at the ceiling, "or maybe it's something more."

Adam eyed Bennett. "Like what?"

Bennett shrugged. "Someone out there may have it in for us—you know, thinking we're responsible for the whale killings. It'd have to be someone local to know the connection between Pequod and the NBWHS. So they may be taunting us, or taking up the defense of the whales, maybe seeking revenge, or planting evidence that makes a case against us stronger—"

"Evidence?" Collin barked. "It's a freakin' file, it's a bunch of electronic bits."

"I'm not talking about physical evidence, just circumstantial."

"Well, that's an incredible reach, but just the same, you have to agree, John was right about the NBWHS going public in condemning the killings."

Bennett thought back to John's letter on the Op-Ed page and had to agree, it was the right thing to do.

Blocks away, the window painted black with the pink pig in the middle of it vibrated with the muffled bass from the loudspeakers inside. The entrance door swung wide as a couple stumbled out and the music, blaring inside, escaped out into the streets in great detail. From down the sidewalk, gradually shedding the shadows, a tall, thin woman took shape, her bouncy brunette hair straight and short. A knee length, black leather jacket, protected her from the chilled night. She entered before the door swung shut and perched herself on a barstool at the end of the bar. Patrick, pulling himself away from the excited chatter of a drunken patron, looked in her direction as he turned to grab a clean glass. He hesitated as he rubbed a towel over the glass and turned back to face the woman. She smiled warmly. Something about her occupied him, and in the moment they exchanged stares, he attempted to pull whatever he had from memory to make sense of the attraction, within the allotted time the stare had before it expired to awkwardness. A name crawled from the depths of his memory.

"Hoosier?" he yelled, his voice filled with expectation. He moved in the woman's direction.

She laughed. "Patrick, how did you guess? It's been so many years."

"I need a beer here!" someone yelled.

For Patrick the room went silent, and his vision collapsed to include only the end of bar where she sat. He tore his apron from his waist and exited through the swing door at the bar. The two embraced.

"You look great," Patrick said, his voice high-pitched.

Hoosier nodded. "You look great too, Patrick."

The two looked on at each other in silence.

Patrick shook his fixation free. "I can join you for a few moments. It's my bar. I call the shots here. No pun intended."

Hoosier giggled. "That would be nice."

"I need a beer here!" the same fellow yelled.

"Excuse me," Patrick said to Hooiser in a calm tone. He turned and cupped his hands around his mouth. "Hey Karl, you wanna give me a hand!"

Karl poked his head out from the office door. Patrick waved him over and pointed to the complaining patron. Karl acknowledged the request.

"So what have you been up to, I mean after the move," Patrick asked.

Hoosier bounced her head from side to side and giggled. "A lot. Let's see how long ago did I move—"

"Twelve years ago I'd say."

"That long ago?"

"Yeah. That was our last time together as counselors at summer camp. Remember those days? They were wild, huh? You, Jeanette, Albert, me, smoking pot under that beached raft. That raft was huge, must have been 24 by 24 or so, a freakin' barge."

Hoosier laughed and tilted her head back. "Yeah, summer camp brings back a lot of memories—you and I and Bennett. I sure hope you and Bennett have patched things up."

Patrick dropped his smile. "Nothing to patch up. He's gone his way and I've gone mine. I think its called growing up."

"Oh?"

"It's a long story." Patrick looked down at the floor, rubbed his forehead, and quickly raised his head. "So what's been going on—what have you been up to since you left that summer so long ago?"

"Let's see...I went off to college. Majored in Marketing."

"Marketing, huh?"

"Yeah well, it sounded good at the time. It actually worked out though, I'm doing pretty well."

"Still living in Indiana? The Hoosier state."

"Nope. Fairfax, Virginia."

Patrick looked at her hands. "So, you married? I don't see a ring."

Hoosier rolled her eyes. "Let's not talk about that—"

"Divorced?"

Hoosier pointed at Patrick. "Bingo! Just one of those things—you know, one of those the other woman things."

Patrick nodded and smiled. "Couldn't keep it in his pants."

"I see you've got a ring?" she continued.

Patrick pulled one side of his mouth with a finger to feign getting hooked. "Yep, ten years ago. I waited two years for you. Finally had to give up." He sneered and laughed.

"Yeah, right. So you happy?"

"Yeah, if you call happy, hitting yourself over the head with a two-by-four every morning. I'd just as soon hit my thumb with a hammer. Let's talk about something else."

"Like what?"

71

"Like what brings you here?"

"I'm volunteering."

"For what?"

"To be a counselor this summer."

"Counselor? You must be losing it."

Hoosier dropped her voice. "I feel like I'm drifting. I've made some bad choices. Right now things are a mess. I decided I needed to get back to my roots. I guess you could say I'm swimming back to a time that brought me great comfort and security. The Island physically, is my oasis figuratively." Hoosier shook her head. "Listen to me, it must be getting late, I'm getting philosophical and I haven't even had a drink—"

"You've come to the right place."

"Thanks, but no thanks. Anyway, I'm here until the fall."

"So where are you staying."

"Nowhere yet. I'm out here looking for a summer rental."

"I can help you with that."

"Good. I'm looking forward to this."

"A counselor, jeez, you can do better than that. Let me take care of you. I can take care of you."

"Thanks, but I want to do this my way. All I want to do is to be able to think and dream like I used to. This is the place to do that."

Patrick shook his head. "How did you know to find me here?"

"I have my ways."

Patrick reached out and grabbed Hoosier's hand. "I'm glad you're here. I'm the guy for you. I hope I'll get to see a lot of you."

"You probably will, more than you ever wanted. Besides I need to hear your tales. But not tonight, I just drove in—"

"Need a place to stay?"

"I've got a room at the Fairfield."

"When can I see you?"

Hoosier stepped down from the stool and nodded. "Maybe you and I and Bennett can get together?"

"We'll get together—you and I."

Hoosier grabbed the door and yelled over the bass, "I'm back!"

At the stop light Bennett pulled the folded white piece of notepad paper from his shirt pocket. On it he had a well scratched-out password to one of his Pequod accounts and directions to Tanya's house—he took one last read to memorize what remained. Two more turns and he would be on her street. He felt his heart race as he took turn number two. His eyes darted, fervently attempting to capture the three-digit street addresses on the houses as he passed them by, jumping from mailbox to porch to over the garage and so on. A black lantern, mounted atop a cylindrical metal post with a cross bar, planted in the middle of the yard, dangled a wooden sign that gilded Tanya's three-digit address. A moat of pink and white petunias surrounded its base. Her yard was a harvest of thick, deep green almost polished grass, well groomed and edged. Two large holly trees stood guard at either side of her front porch. Bennett studied the sign again and pulled the note from his shirt pocket—the addresses matched—the notion of Tanya having a green thumb never would have occurred to him.

He was greeted by a series of heavy footsteps and the muffled voice of a woman from behind the door. Bennett took the moment to check that his fly was up and that his shoes were tied. A small verdigris frog sat in the corner of the cement porch near his feet. He picked it up for a closer inspection. As he rotated the frog he noticed an alien object resting in its mouth. He pinched the object at its visible end and extracted it from the frog. He pulled out a key. Hearing the approach of footsteps, he replaced the key and set the frog down in its original place.

Tanya's ravishing image floated in the door, framed in a transparent surrealistic mixture of sky and flora captured in the glass—the rich, green arbor of her yard forming a lush garland about her auburn hair. Bennett waved awkwardly. Tanya waved him in, and after he stumbled, she grabbed him and offered him a tour of the house. As he followed from room to room he noted each room sparingly inhabited by an assortment of a few pieces of furniture and pictures, with the exception of the extra bedroom, which threw a busy exercise machine into the mix. The machine consisted of wood, pulleys, steel, and weights—an elaborate concoction that left Bennett scratching his head. In the end, the tour did not bring about the warm feeling he had expected. Perhaps in time—after all she had just moved in and making a house a home required time—a melding of the owner and the structure, not unlike a dog taking on the face of its human family over time. Upstairs, at the end of the hall, Tanya brought Bennett into her study.

"There it is…" She presented her computer to Bennett by directing the open palm of her hand at the device, like a model would an elaborate showroom product up for sale.

"Wow!" Bennett said automatically, looking astonished, just as he had earlier when she had shown him her hanging baskets in the back, and the grandfather clock in the living room, and the Mickey turnabout cookie jar in the kitchen, and everything else she had pointed out in between.

"So, it won't be too much trouble to hook up?" she asked, sheepishly.

"Not at all. I'm at your service, Ma'am," Bennett said, saluting.

He began to disassemble the three cardboard boxes on the floor, removing the foam protectors that enclosed the hardware inside. A composite, wood veneered desk was jutted against the wall, the only piece of furniture in the room. He astutely concluded the machine should be setup on it.

"So, you know a lot about computers, huh?" Tanya asked, resting against the wall opposite the desk.

Bennett pulled open the big box. "Oh, a few things you could say." He did know a lot. Some of it—though consciously forgotten, going back as far as the stuff where one had to hookup an external cassette tape machine to provide persistent storage of data—still sat in a coat of dust on one of the tables in the laboratory of his mind, the part of his mind that stored tinkering-awareness, the part he instinctively drew upon, without a conscious thought of what all was in there, only that the stream provided him clues to untangle new and unexplored issues. Strewn throughout the laboratory were radios, cassette recorders, loud speakers, alarm clocks, windup devices, too many more to list—all of which had either been dismantled cautiously by a screwdriver, or eviscerated by cruder tools of exploration, such as the knife used to disembowel one of his mother's dolls to retrieve the mechanical contents inside. Of course in that same laboratory, planted in with the rest of it, the remains of his grandfather's timepiece, the gold one with the fancy arms, one of which, his tinkering had snapped off. But all progress must be made at a price. He felt guilty just the same.

"You're so modest—"

"Not really, there are guys out there who can run circles around me." Bennett lifted the tower from the box and placed it on the floor beside the desk, the smell of the new equipment wafted from his actions. He liked the smell.

"I seriously doubt that."

Bennett began pulling the sub-boxes from the big box. From the first one he pulled out the keyboard. "Well, there's a guy out there right now who's toying with me. Sending me encrypted stuff, seeing if I can crack it and get to the prize."

Tanya snickered. "What is it, a box of Cracker Jacks?"

"You may be onto something. I've wondered myself why I've been so motivated by this pursuit and now maybe you've hit the nail on the head, seeing as when I was a kid I was consumed by the mystery of the Cracker Jack prize. I used to beg my dad to buy them for me, Mom never would, but I could sway Dad."

"I'll have to meet him someday. He's probably a lot like my dad. You know what I'm saying."

Bennett stopped what he was doing. "He died a few years back."

74

"I'm sorry—"

"It's okay, it was a while ago."

"Still, it must be difficult at times. Did he die of natural—"

"Boy, this is a nice machine you have here." Bennett hooked up the mouse and keyboard.

Tanya nodded, as if tacitly acknowledging his right to not have to discuss the subject. "I'll bet you figure it out." She leaned over and looked at her legs.

"What out?"

"The encrypted puzzle."

"Unfortunately it's not as easy as eating a pound of caramel corn to get to the answer. But you're right, I might, years from now, when processing power has caught up with today's key lengths."

"I'll give you a couple of weeks."

"I don't know…"

"Dinner?"

"Huh?" Bennett looked up from the big box.

"We'll bet dinner."

Bennett grinned. "It's a deal." He couldn't lose on this bet. He finished pulling the machine together and flipped the switch. Text darted down the screen.

"Is it working?"

Bennett nodded. "Yep, it's just booting up. In a few minutes we'll have you up and running."

Tanya clapped her hands. "Good show, Bennett."

"Glad to be of help. But we're not done yet."

"What's left?"

"We need to get you connected to the Internet."

Tanya knitted her brow. "Why?"

Wide eyed Bennet asked, "Why?" "Everybody needs the Internet, it's become an integral part of everyday human life. It's more important than a phone. These days everyone is connected to the Internet."

"Is that a Pequod marketing speech?"

"So you've heard our ad on the radio?"

Tanya laughed. "Just giving you a hard time."

Bennett got a poor-me look. "I'm used to it."

Tanya sighed. "Seriously though, I am concerned about getting a virus and all that other malicious creepy stuff crawling the web."

"Creepy stuff?"

"You know, porn, the weirdos into it, virtual stalking—"

"What are you talking about?"

"I had a friend who met this guy in a chat room. They started out on friendly terms and then one day he went weird on her. Started sending her naked pictures of himself. She changed her address, but he found her. She tried to change her Internet identity three times, and all three times he found her. She had to stop surfing completely."

Bennett shook his head. "I don't think that happens very often—"

"You could even have some of those weirdos at Pequod."

"Man, are you going to suggest every organization I affiliate with could have weirdos?"

Tanya winked. "Perhaps…"

"I know everyone at Pequod, and sure we have our share of eccentrics, but nobody on the scale you're talking about." He thought of the office, about Mary and her doghouse, complete with a Yorkshire Terrier inside, and James, the blowup alien floating over his cubicle like some menacing dirigible, and Sam with her pet tarantula in the fishbowl, the one that had escaped twice.

"You never know who might have a dark side. I might have a dark side," Tanya said, attempting a Russian accent.

Bennett chuckled and shook his head. "Yeah, right."

Tanya pursed her lips and swished them from side to side. "Are you sure I won't get any weird stuff?"

Bennett laughed. "We've got screening software that will protect you. I can almost guarantee it."

Tanya's eye's popped wide. "Almost?"

"Alright, forget the 'almost' part."

She sighed, her index finger winding her auburn hair around it. "Okay, sign me up."

Bennett reached Adam on the phone and had the account setup in a matter of minutes. He plugged the internal modem from Tanya's machine into a phone jack in the wall.

"You're in business," he said, as the whining of the modem ceased and the computer indicated the machine was logging onto the Pequod network. "Where do you want to go?"

Tanya squinted and looked at the display. "Where's my spreadsheet software?"

Bennett chuckled under his breath. "You've got all of cyber world out there and you want to look at your local spreadsheet software?"

"Can I inventory my stuff on the Internet?"

"Possibly, but I wouldn't."

"Then let's bring up the spreadsheet software."

Bennett laughed and shook his head in an exaggerated fashion. "Okay, you win."

The rain hammered on the roof of DeWald's automobile. The wipers, despite a setting at their highest mark, could not keep pace with the hard driving rain. He slowed considerably, grabbing a napkin from the glove box and rubbing the film of condensation from his side window. As he passed the three-story dwellings lining the street, he searched for the Mendosa house. Of all the parts that made up his job, this was the one he disliked the most. There was no easy way to question family members of the deceased. With little deviation, he found a strong desire on their part to bring the guilty to justice, but when it came to extracting the information necessary to do just that, the pain associated with it sometimes hindered their efforts to such a degree that questioning them further became all but impossible. Nobody likes to peck at a wound, with the exception of a dentist. But a good investigator has to, and sometimes that means probing the most painful areas.

The pale blue house with dark blue shutters appeared in the circle cleared in the window. He pulled to the curb to park. The information he had gleaned from the transcript of Thomas' earlier interview with the Mendosa family had proved unsatisfying to him. There were key areas that Thomas, in his greenness, had overlooked. DeWald took a moment to flip through his notes. Cynthia Mendosa was the second victim. A college student, she had come down for the weekend to spend it with her parents. She had left the house at 7:00 p.m. on a Saturday to visit a local ice cream shop to pickup her and her mother's favorite ice cream. She never returned. Her car was found abandoned, far away at Popes Island, a patch of land along the highway between New Bedford and the nearby town of Fairhaven.

The car door opened to a veritable river rushing down the curb. DeWald stepped over it into the yard. At the base of a street lamp near the curb were votive candles and a weathered looking wreath made of silk flowers with the deceased's name on it. DeWald recalled watching a vigil organized by the neighborhood on television the day after they had discovered Cynthia missing. He picked up a candle that had been forced to the curb by the wind and placed it back by the wreath.

A plump woman wearing an apron answered the door. A wonderful scent flowed out when she opened it.

"It's linguisa," she said after DeWald inquired. "Will you join us for supper?"

DeWald tactfully declined.

"Have you learned anything—about Cynthia's death I mean?" Mrs. Mendosa asked, her face garnering expectation, as she seated herself and pointed to the sofa for DeWald to seat himself.

Douglas Haydon

DeWald looked over Mrs. Mendosa's shoulder to a photograph hanging on the wall, and saw a beautiful olive skinned girl wearing a purple high school cap and gown, her brown eyes lunging at the future, subdued only by her soft smile. His own niece came to mind, though not old enough to be graduating from high school, he fully expected her to, and wasn't that what it was all about, expectations, and what could be worse than to have them cut short...He sighed. "I'm afraid we haven't learned a whole lot more."

Mrs. Mendosa nodded, saying nothing.

"But that's one of the reasons I'm here. I have some further questions I'd like to run by you if that's okay?"

"Of course. I will do anything to help Cynthia."

DeWald pulled his notes from his pocket. "I really appreciate your taking the time. I won't stay long." He scanned his notes. "I know Detective Thomas asked you about Cynthia's friends and acquaintances and so forth, but I want to try and expand the search further."

Mrs. Mendosa wrinkled her forehead. "What do you mean?"

"How did she spend her time? What places did she frequent? Did she have hobbies? Did she go to the beach? Those sorts of things."

Mrs. Mendosa tilted her head back and rubbed her chin with her fingers. "Well, she liked to run—"

"Where did she run?"

"Well, I don't know where up at the school, but when she's here at home she liked to run in this neighborhood."

"Never anywhere else?"

"Not really."

"Do you know the specific route she took through the neighborhood?"

"Roughly. I know where I've spotted her when out driving." A tear formed in the corner of her eye. "Sometimes I think I still see her..."

DeWald nodded and paused, giving Mrs. Mendosa a moment. "Could you draw me a map later?"

"Sure."

"Was she into any special organizations?"

"Organizations?"

"Like Save the Whales, or something like that..."

She focused on the ceiling as if she might find some lost memory floating up there. She lowered her eyes and shook her head.

"What else did she do?"

Mrs. Mendosa raised an eye. "She worked awfully hard on her school work."

"Okay..."

78

"She spent a lot of time on her PC—"

"Doing what? Do you know?"

"I think she called it chatting?"

"Hmmm, did she mention any interesting characters she 'chatted' with?"

Mrs. Mendosa shook her head.

"Would you mind if I had someone look at her PC?"

"Not at all, but Mr. Thomas already went through it."

"I know he did. Unfortunately Mr. Thomas is not adequately trained in computer forensics. I would like to get an expert to take a look at the box."

DeWald noticed a worried look on Mrs. Mendosa's face. "Is something bothering you, Mrs. Mendosa?"

"No, it's just that I may have erased something important on her machine. I really don't know anything about them, but when she disappeared I printed out the missing posters from her PC."

"Don't worry. No harm done there. So what else can you tell me?"

"She played tennis."

"Where?"

"Over at Fairhaven High School."

"Anything else?"

Mrs. Mendosa smiled. "I have a lifetime of memories I could tell you, but I'm sure it would be of no benefit to your investigation. She was, and always will be a beautiful child."

DeWald nodded and felt himself drawn back to the photograph. The child's beautiful brown eyes beseeched him. Twenty-one years taken. Twenty-one years pulled out from under a family. DeWald felt an emotional wince—time for a painful question. Unlike a dentist, he had no Novocain.

"Mrs. Mendosa, forgive me, but lab results indicate that your daughter had traces of marijuana in her blood. Do you know where, or from whom, she might have purchased the drug?

She scrunched her face and slowly shook her head. "There must be some mistake, my daughter would never take drugs."

"It is possible that the lab made a mistake—"

"Yes, yes, they must have."

"Did she ever hang out with what you might consider 'questionable' characters?"

Her jaw relaxed and her eyes focused to the floor. "No, she was always with very decent people."

No need to pursue it any further. He asked a few more questions and jotted down notes. Mrs. Mendosa drew a rough map of where she thought Cynthia might have jogged. DeWald finished and stepped back out into the rain. In his hand he clutched a warm brown bag trailing steam as he walked—a linquisa sandwich. Smelled awful good. Mrs. Mendosa had insisted.

The small chop lapped the side of Patrick's Whaler. The Envinrudes were silent. Up in the midnight sky, the moon shined as a hairline crescent. Paul squatted at the bow, maneuvering the spotlight over the surface of the water. Patrick stood behind the wheel wearing headphones. He heard their songs off in the distance. The microphone hanging over the side brought them to him. He found the songs eerie and mysterious, and unlike Paul, understood they actually were communicating something. Paul told him the sound was nothing more than the whales having a group fart.

Patrick looked back towards the shore, now nothing more than a few spots of light from powerful sources. In the distance a marker buoy clanged its bell in solitude. A capricious breeze swept across the boat. Patrick slid back the panel for the storage compartment built into the steering column. The faint light from the moon reflected on the chrome barrel of Karl's handgun. He inched the panel back snug to close the compartment. He could hear the sound of outboards approaching.

Squelching the breeze, there was a loud *pop* and the sound of something shearing the air.

"What the hell was that?" Paul asked.

Patrick did a quick 360. He whispered, "I think somebody shot on us. Get down."

"Identify yourselves!" Came a bodiless voice from the water over to the left side of the boat.

"Coast Guard?" Paul asked in a whisper.

"Could be."

"What do we do?"

Patrick felt a strong urge to flee. "Maybe run for it?"

"Start the engines, man."

Patrick grabbed the gun from behind the panel and held it at his side. He looked up and scanned the water once again. To his dismay, he found they were encircled by four or more privately owned powerboats. "We're surrounded."

"By whom?"

"I don't know. But its not the Coast Guard or Harbor Patrol."

"What are you boys up to?" The same voice rang from the left, now much closer.

Patrick stood up. "Fishing. What's up?"

A bearded man held a megaphone. He dropped the megaphone to his side. "Fishing at this hour?"

"Now is the best time."

The bearded man looked up and down the boat. "What are you fishing for?"

"Blues, strippers, chicks. Whatever is biting."

"Chicks?"

Patrick whispered to Paul, "we've got a real genius here." He turned to face the man and raised his voice. "Never mind. It's a joke."

"Oh." A pause. "So what's with the spotlight?"

Patrick let loose a deep breathe. "Alright, you're onto us, we're searching for those dudes killing the whales."

"Really?"

"Yeah, swear to it."

"Shoot, thought we had something."

"If you don't mind me asking, what are you boys up to?"

"Searching for the same guys you are."

"How do I know you guys aren't the ones hunting the whales?"

The bearded man laughed. "Hell man, we're just a bunch of drunks trying to make some extra cash. Do you think we're capable of hunting whales?" Laughter burst out from all the boats. "Hell, why don't you join us? We can team. We split equally between each man. What do you say?"

"We've been out for a while. We we're just about to head it in. Otherwise we'd join you."

"You wouldn't be on to something you didn't want to share with us, would you?"

"Hell man, we'd need all the help we could get if we did."

"Maybe next time then."

"Yeah, but do me a favor, ask questions first then shoot."

"Won't happen again. That was Bill over there; he's a little trigger-happy. Especially after a few brews."

"Good luck fellows," Patrick said, and waved.

The boats shifted from neutral and tore out to the east.

"Ahab is paying us more money," Patrick said, watching the boats disappear. "Much more money."

"Should we head in?" Paul asked, shaken by the encounter.

"I think they'll be long gone, besides, it takes too much time to get out at this range, I don't want to waste the effort."

Paul looked in the direction the boats took off in. Patrick observed as Paul listened intently. A firecracker, if it were exploded at the moment, would have given the man a heart attack.

A half-hour later, Paul turned his eyes away from the direction of the departed boats. In the spotlight Patrick could see the details of his face, specifically the red fibers, the ones caught in the bristles of his beard, that cast a crimson shadow on his cheeks, that looked like dried blood, the ones from the red wool blanket he wrapped himself in each night as he drifted off to sleep on the sofa in his mother's home. He could picture him there, because often he found him there, reeking of alcohol, his boots still on his feet, an open bag of chips, like a cornucopia, flowing chips out onto the floor, a porn magazine with the foldout flopping over the edge of the coffee table. He could picture the old woman, watering her plants in the kitchen, not knowing what to do with the lout passed out on her sofa.

"You ever been to Wally's?" Paul asked, shifting from a squat to seating himself Indian style.

"What?" Patrick asked, removing the headphones.

"You ever been to Wally's?"

"No, why?" Patrick put the headphones back on, but removed the headphone over his left ear.

"Best damn hot dog in town."

"Is that right?"

"Fresh sauerkraut, yellow mustard, onions, jalapenos—"

"Hot peppers?"

"Yeah, makes a hellofa wiener, let me tell you."

"You had Porgy cake?"

"What's that?"

"Chocolate cake with the marshmallow filling. You can get them at 7-Eleven. When Karl and I had the munchies the other night we downed an even dozen. You don't want to go over a dozen with those things though. Nasty."

"Nothing tames the munchies like salt and vinegar chips. You had those?"

"My granddad lives on them. Always has them at the house. When we were young, Bennett and I would sneak the bag outside and eat most of it down at the beach. Granddad never cared. Just went out and bought another one, that and six or seven pies. I don't think he ever saw a pie he didn't like. He would beg Bennett and I to take some of them off his hands after he realized how much he had over bought."

"Your Granddad's cool, I really like him."

"Yeah, I'm gonna have to pay him a visit pretty soon."

Both men fell silent for a moment.

"So how's Karen? You still seeing her? She still eating a bag of Doritos everyday?" Patrick asked.

"Hell no. The wench gave me lip. I don't take that off of no woman."

"Charming, Paul."

"Screw that. A woman's only good for one thing and green will do all the seducing you need, nobody needs charm."

"So, you're on your own, huh?"

"Sure am."

"Man, you could fall off the face of this earth and nobody would be the wiser."

"Damn straight."

"You could fall off this boat and drown and nobody would even come looking for you."

"Well, I hope you would wonder where I went."

Patrick rolled his eyes. "No, I'd just sit here oblivious. Come on, Paul."

After a moment of silence, Paul scrunched his face. "So where are you going with this conversation?"

"Passing time, Paul, passing the time. If this was any more fascinating I'd blackout."

"We'll let's change the subject. Okay?"

Patrick replaced the headphone to his left ear. Paul turned his ruddy beard around to face directly toward the bow. It would be almost a humane action, Patrick thought. A man whose life subsisted on booze, women, and food—what kind of a life was that? Sure, it was all Paul had, but sometimes the end of one's life is welcomed, when life itself is so puny, narrowed to a tiny feeble spec of experience, and no amount of compassion can possibly pry it so that it might expand wider. Why not set him free?

And I need the money. I've got something good going here. Something I can't afford to have him screw up.

Patrick slid back the panel. Curiously his pulse increased, and though he had not yet forced himself to think about it, it occurred to him his mind was preparing itself, mixing the right chemicals at the right consistency to forge a reaction powerful enough to extort his temperate personality to pull a trigger. The gun glistened in the moonlight as his hand reached for it. The thought of killing somebody and knowing that it might really happen brought the seas, the boat, his hand, into a surrealistic aura, a higher level of experience. He felt his mind failing to allow itself to contemplate the consequences, blocking them, much as it enlisted endorphins to obscure physical pain. He had an urge, consisting of a compelling desire to shoot, to watch Paul clutch his side and then tumble from the boat to the sea.

A friend, kill a friend?

He released his grip on the weapon. His breathing was considerably strained. Sweat dripped from his upper lip and off his chin. There was no choice; Paul would talk. He watched as a spectator at a distance, experiencing something out of body, as his hand once again gripped the handle of the gun. In horror the weapon was pulled from the storage area and aimed at Paul's back. The weapon was cocked. He licked the sweat from his lips. His hand shook uncontrollably—

No—can't do it. Not now.

The weapon dropped to his side.

"What about Cordoza's Pizza. Ain't that the best pizza you've ever had?" Paul asked.

The sun burst through the windshield like a halogen lamp as Bennett leaned into the back seat of his car to retrieve his briefcase. He'd been thinking about Tanya and what would be his next step, so he performed the action mechanically. After more than a few seconds of his hand fluttering in the space behind the passenger seat, and failing to encounter his briefcase, he paid the effort its due attention. He found it on the back seat, not in its standard position on the floorboard, and also was overcome by a feeling that something more was not quite right. He shook his head and got out of his car. As he was ascending the stairs at Pequod it occurred to him what had changed: his green rubber boots were no longer piled on the floorboard. He returned to his car and did a thorough check of its contents to ensure himself that something else wasn't missing. As best he could recall everything was still there, minus the boots.

He found Collin in the kitchen area stirring instant oatmeal into a bowl of boiling water he had heated in the microwave. The scent of cinnamon and apple drifted in the steam billowing from the creation. When he informed Collin of the break-in, Collin showed little concern and even went so far as to doubt a break-in had actually taken place. His rationale following the lines that nobody would risk breaking into a car for a well-worn pair of boots—car stereos, CDs, radar detectors, yes—boots, no. He asked Bennett if he planned on contacting the police, something Bennett shrugged off as a waste of time considering his loss totaled no more than a few dollars. Collin only nodded, having slurped down a spoonful of oatmeal. As Bennett began to leave, having fixed his coffee, Collin spoke with a mouthful, informing him that Sandra wished him to call her. Bennett nodded, at first ignoring the trail of oatmeal dripping from the corner of Collin's mouth. But then: "Do something about that oatmeal—*please*."

Back at his office, Bennett got Sandra on the third ring.

"Interesting stuff you've got here," she said.

"How so?" Bennett began doodling out a whale with a Pequod ink pen on an old memo.

"Well first it's a hybrid of encryption algorithms."

"Really? How can you tell?"

"I actually decrypted parts of it."

Bennett drew water coming out of the whale's spout. "Anything meaningful?"

"I think so. It's just a guess at this point, but I believe part of the payload for this file is actually an image."

"Why do you think that?"

"I decrypted just enough to get an indication of a signature for a common image format."

"Hmmm…that's interesting."

"Yeah, but it's not what I find most interesting."

"What's that?"

"The part that I was able to decrypt used some variations—albeit good variations—on some old tricks that have been around for years. It's a mixed bag of bit shifting and *xor* stuff."

"Even I'm familiar with that—"

"And that's what's so interesting about it. Why would someone use common archaic stuff—a good choice maybe ten years ago—when there's free stuff out there available to anyone, and it's 100 times better?"

"You've got me. Maybe it's some old timer who refuses to go with the times." The theory of a young punk was appearing less plausible.

"Possible, but somehow I don't think so."

"Do you think you can crack the rest of it?"

"Well I've kind of hit a brick wall. Although the core algorithms are basic, the seemingly random pattern of their engagement makes things difficult. So I guess what I'm saying is I'm not sure if I have the time to attack this like it should be attacked. Is this really critical?"

Bennett told her not to bother, that he'd take over from here. He thanked her for the information. It was very useful. An Image? *An Image*? What could it be an image of? Anything. Perhaps Collin was right—perhaps it was an image of the word *Gotcha*. He scribbled more on the whale. He winced. The drawing brought back memories of his days in grammar school. At every opportunity to participate in arts and crafts, and he had many through all six grades, he always created a whale. If it was watercolor he painted a whale, if it was clay, he molded a whale, if it was crayons, he drew a whale. In the teacher's lounge, as he later found out, he was known as the whale kid. Bennett took his pen and scribbled over the drawing until it became a solid block of blue ink.

The morning paper banged loudly against the front door, Bennett jumped and nearly severed his neck with his razor. Damn paperboy. He finished shaving and wiped his face clean. He dressed and opened all his windows. The morning was cool and the air was fragrant. The scent of Honeysuckle drifted through the screen in his kitchen window. He breathed it in deeply. The scent always reminded him of Granddad's cottage on the Cape—honeysuckle draped the wooden fence in the backyard. He poured himself a cup of coffee and sprawled the paper across his kitchen table. Nothing new on the strangler; pressures were mounting on the Police Department to make more progress; the editorial page was thick with voices from concerned citizens.

Ezra Chaplin was dead.

Bennett's jaw slackened. He remembered Ezra from grammar school. The token overweight kid all the other school children relentlessly teased. The only exception to this was Bennett, who left him alone, which caused Ezra to cling to him. Whenever Ezra spotted Granddad's car in the drive of Bennett's house, he knew, barring bad weather, a fishing trip was an automatic, so he would stop by begging to join Bennett and his granddad on their trip to the causeway. Though he hadn't seen him for years, the news saddened him. The picture in the paper showed a chubby face with a bushy beard. The article stated Ezra was victim to an accidental shooting. He had been part of a group on a bounty-hunting expedition for the whale killers. Bennett found himself staring at the photograph. Who would have known, so many years ago as they played kickball in the schoolyard that Ezra would die this way?

Bennett picked up the phone and dialed Granddad. It rang five times before and out-of-breath voice answered from the other end.

"Just got back from fishing," Granddad explained. "Amazing, I was just about to call you."

Bennett brought him up to date on what had been going on and filled him in on Ezra.

"That's a real shame," Granddad said, "as I recall he was a good kid." A cough.

He could picture his Granddad on the other end of the phone: pea green waders, a Pabst Blue Ribbon ball cap, white hair like thin bleached straw forking out the back of the cap, his heavy brow weeping from the mist of the sea. "Remember the time he got the hook caught in his pants?"

"Yeah, sure do, was quite the fiasco…"

Bennett looked over at the photograph of his granddad resting on his end table. He loved Granddad's smile. It swallowed all of life's difficulties. "I'll have to pay you a visit sometime soon."

"As a matter of fact, that's what I was gonna call you about. To have you over." Another cough.

"That'd be great. Are you okay?"

"A little congested. I'm inviting Patrick too…"

Bennett dropped the phone to his side and shook his head. He heard his Granddad's voice cackle through the receiver and returned it to his ear. "I know what you're trying to do, Granddad, and it's something I've approached many times myself, but he doesn't want to be around me—"

"But he has to put all that behind him."

"He will always blame me, Granddad. I can't change that. Believe me, if I could go back in time—"

"No, let's not head in that direction…" Bennett heard a sigh. "You're handling it all very well. It's Patrick that isn't."

"Not on some days. But the thing is, whenever Patrick and I are together we can't seem to avoid the subject. We're like moths to a flame."

"It just kills me to see you two not getting along. You used to be such great buddies, you did everything together."

"But we grew-up, Granddad." At least one of us, Bennett thought.

"It was that Paul fellow that started it all. He was a bad influence."

"Patrick is Patrick, all by himself."

"That Paul was born bad. Some of them are you know. Bad from the start, and nothing that happens in their lives is gonna change them."

"I don't care for Paul either, but there's nothing we can do about that." Bennett took a long sip from his coffee and combed his fingers through his hair. "When do you want to get together?"

"So you'll give it a try?"

"I'll try." And it will be more difficult than you can imagine, Granddad.

"Good, Bennett. I thought this weekend. We could have a cookout. Patrick can make those fantastic burgers of his."

"I'll stop by McDonalds on the way," Bennett said as an indirect insult.

"What's that?"

He was sure the comment raised a questioning brow and removed the hat from his granddad's head and it was now held at the side of the waders. "Can I bring a friend?"

"Why sure, who is it?"

"A girl I met."

"Ah…You struck it rich, huh? What's her name?"

"Tanya…"

"Tanya, such a pretty name. I can't wait to meet her."

Bennett heard a car pull up to the curb in front of his home. A brunette climbed out of the car and began her way up the front sidewalk. Bennett squinted. "Who in the heck is this?"

"Huh?"

"Some lady is walking up towards the house."

"Probably a solicitor of some sort—"

"It's Hoosier!"

"Hoosier?"

"Yes, you remember Hoosier."

"Hoosier?"

"Got to go Granddad, see you this weekend—"

Patrick dealt himself and Karl five cards each. He picked up his cards and began arranging them. His eyes glanced down at the folded newspaper in the seat beside him. The headline caught his attention earlier in the day: *Arctic Sunrise Sets Course For New Bedford*. The boat in the photo looked daunting. The article stated that with agreeable weather conditions, the boat could arrive in as early as a couple of weeks.

Uncle Jack respected whales. During one fishing excursion in particular when they encountered a herd, Uncle Jack gave them a wide berth, explaining we owed the whale a great deal of respect.

"Why do we play this damn card game everyday?" Karl asked, leaning back in his chair.

"Because you love it so much."

"Oh yeah, I forgot. Thanks for reminding me."

"Anytime Karl, anytime." During that fishing trip, he, along with Bennett and Uncle Jack watched the whales for more than an hour, Uncle Jack explaining all he knew about the creatures—what made them so special. And now he, Patrick, killed them, killed them for money. These massive living creatures, of great complexity, traded for a few thousand dollars. And it was he who made the decision when these marvelous beasts would be pulled from their watery domain to die. God-like. He wore a reputation for killing stuff. Some of it earned, some of it not, part of the reputation his own fault.

Karl placed two cards on his side of the cribbage board.

"Excuse me," Patrick said, "who dealt?"

"You."

"Who gets the crib then?"

"You."

Patrick frowned and pointed at the two cards Karl had just put down. Karl sighed and moved the cards to Patrick's side of the cribbage board.

"They killed the guy Paul and I ran into last night fishing," Patrick said, tossing two of his own cards on top of Karl's.

"What guy is that?"

"He was part of a group of vigilantes looking for those guys killing the whales."

"They came after you?"

"They thought we were the guys, can you believe it?"

"You and Paul?" Karl burst out laughing. "You two couldn't land a *porgy*!"

"Just fifty pound strippers and blues—only the best."

"In your dreams!"

Karl made thirty-one and advanced his peg two holes. "Now ain't that something."

"What?"

"Those fellows thinking you're whale killers," Karl said as he fell into another burst of uncontrollable laughter.

A loud rap came from the front door. Patrick and Karl eyed one another, both looking surprised. A second round of rapping came from the door. Patrick slid his chair back, got up, and walked over to the pink pig on the window. He peered through a tiny clear portion in the pig's eye. Outside, he saw a man in a suit looking at the door.

"Who is it?" whispered Karl.

Patrick shook his head.

Karl held his palm up to Patrick as if to say wait. He then skirted into the back room. He returned quickly looking concerned. "I can't find my gun," he whispered. "What's going on here?"

"We don't need it," Patrick said scornfully. "Let's just let the guy in—it's no big deal."

Patrick moved to the door and yelled out, "Just a minute." Removing the deadbolt he pulled the door open. He observed a Hollywood looking suit with a young punk in it.

"Detective Thomas." Thomas extended out his hand. Patrick shook it looking puzzled. "I'm looking for a Patrick Arnold."

Without releasing his grip, Patrick said, "That'd be me. What's up?"

"Need to ask a few questions. Got a moment?" Thomas wrinkled his nose and looked beyond Patrick.

"No problem," Patrick said, taking a nearby seat. He motioned Thomas to a chair across the table. "No problem at all."

Thomas waited for his eyes to adjust to the low lighting and gave the room a once over, chewing his gum methodically. He pulled a pen and pad of paper from his inner suit pocket. "Can I ask your whereabouts at 4:00 am this morning?"

Patrick shrugged. "Out fishing with my friend Paul."

"Is that all?"

"What do you mean *all*? Did I eat a donut and take a piss off the bow—yes."

"Cute. Want to continue this behind bars? Keep it up and you will."

Patrick shrugged. "Here's fine. Next question?"

Thomas staring directly into Patrick's eyes: "I have word from some folks who came upon your vessel—the *Where II*, correct?"

"Yes."

"That you and your friend were out bounty hunting for the whale killers."

Patrick shifted in his chair. "Killing two birds with one stone. I was keeping an eye out just in case."

"Do you fish late at night often?"

Patrick laughed. "All the time, it's the best time to fish."

"Guys at the docks say you never come back with anything..." Thomas leaned back in his chair. He pulled a cigarette from his pocket.

"No smoking."

"In a bar?"

"It's off hours."

"So how do you explain it? All night fishing and nothing to show for it. No blues, no strippers, no scup, not even a porgy."

Patrick looked over to Karl. "We're lousy fisherman."

"Your fishing partner, Paul, what does he do for a living?"

"A shop teacher at the local high school."

"Is he good with wood?"

Patrick grinned and shook his head. "I don't know. What does that have to do with anything?"

"Probably nothing."

"Have I done something wrong?"

"Nope. As long as you're just fishing of course."

"Just fishing," Patrick said, emphatically.

Thomas slowly nodded his head. Looked around once more and got up from the table. "I thank you for your time, Mr. Arnold."

Patrick got up from his chair. "Anytime."

Thomas moved towards the door and Patrick followed. "I can find my way out," Thomas said. When he reached the door he turned to face

Patrick. "By the way, do you think sometime I might be able to take a peek at the *Where II*? I hear she's a fast boat."

Patrick shrugged. "Anytime."

"Sweet." Thomas disappeared out the front door.

Patrick turned to Karl. "What a moron."

Karl laughed and walked into the back office muttering, "whale killers" under his breath and shaking his head. Patrick returned to the table and sat back in his chair. He dared not rush out to the *Where II*, as compelling as the desire was, he knew that's what the detective would be looking for him to do. Instead he focused on the blank far wall of the bar and began a mental exercise of going over every inch of his boat from memory. He went over the floorboards, the railings, steering column, cleats, everything, and thought back to the morning of their last return—did they remember to remove the harpoons?

"Hoosier," Bennett yelled out after opening his front door. "Man, how are you?"

Hoosier smiled and wrapped her arms graciously around Bennett. "It's so good to see you. I'm surprised you remember who I am."

"How could I forget my candy-sharing, always-playing-together friend from childhood?" When he thought of Hoosier, as he still did every now and then, the images most conjured up in his mind were those of the Powerhouse, the white painted cinderblock building with the Disney images sketched on the outside walls, the place where they first met, and in the early years of their friendship, the place where they spent the bulk of their time together. He even recalled the pipes, artifacts from a one time functional power source, out back, painted green, made to look like a dragon buried halfway in the earth.

"Puberty," Hoosier said.

"Huh?"

"Friend not only through childhood, but puberty as well, remember?"

"How can I forget? How can any of us forget?"

He poured Hoosier a cup of coffee and joined her on the sofa. She filled him in on why she was in town. As she was finishing, she noted Bennett eyeing the clock on the wall. "Am I keeping you from work?"

Bennett shook his head. "Are you kidding? Work takes second seat to my buddy."

Hoosier nodded. She looked Bennett directly in the eyes. "I had the biggest crush on you."

"Boy, you just jump right into things, don't you?"

91

"I do and I haven't changed a bit, have I?"

"Nope." Bennett grinned sheepishly. "Likewise."

"Likewise what?"

"Likewise I liked you."

"We never really told each other, did we?"

"I guess we just knew, right? But there was that other guy you liked too. Mark what's-his-name."

"Mark Pippen?"

"Yeah."

"Are you serious? The guy thought he was a god; he was a moron." Hoosier chuckled and took a sip of her coffee.

"We'll I know guys were falling over each other to ask you out."

"Yeah, right. The only other guy I remember asking me out was Collin something-or-other. I think they thought you and I were an item—"

The reaction was delayed. "Collin? Collin, asked you out? Did you go out with him?"

"You remember—we talked about it," she said, watching Bennett shake his head. She continued, "Well, I asked for your advice. I turned him down because he was older than me."

"Was that my advice?"

"Probably."

Bennett lifted a brow and shrugged and went for his coffee.

Hoosier laughed. Bennett could see her eyes peering back in time. "And what about Peggy—"

"Peggy Jenkins?"

"Yeah, what a whore!"

Bennett rubbed the side of his face. "It just seems so long ago. It's so funny to think about it now."

"I know...I know." Hoosier glanced over the room. "You've got a nice place here—still into that whale stuff."

"Yeah, well, I guess it's just a part of me." Bennett absorbed the face unseen for so long and was amazed how comfortable it felt to him. "What about you, what have you been up to?"

Hoosier fell back against the sofa and filled Bennett in on her history since she left. "It's amazing how screwed up we can get," she said at the end.

Bennett nodded. "Yeah, I guess somehow we think the way we are at one point in time is the way we will be for all of our life, but we change— sometimes overnight."

"When I left here, I was excited about life, excited about the potential of what I could become. But it's gone now, the excitement that is."

"So you're back to reclaim it, huh?"

"Sounds crazy doesn't it?"

"Nothing sounds crazy anymore."

"So you have anyone special in your life?"

It took some time, probably too much time, but he did manage to think of Tanya.

Hoosier spoke before he could open his mouth. "I'm sorry, Bennett—awful presumptuous of me to ask."

He waved his palm at her, dismissing her apology. "Time will tell. Her name is Tanya."

"Ah, Tanya. The name brings to mind words like daring, risky, adventurous, and perhaps even *spicy*. Like Natasha from Bullwinkle."

Bennett chuckled. "Only you'd have thought of that. I guess that makes me Boris Badenov and I live in Pottsylvania."

"You haven't changed much."

Bennett put his hand on his gut. "A few pounds heavier."

"I didn't mean it that way. You're still that easy-going laid-back guy that I remember."

"Yeah, I suppose."

Hoosier leaned over towards Bennett, kissing him on the cheek. "Thank you for remaining the one constant I've had through all these years."

Bennett felt heat come over his face.

"You still blush too! How wonderful!"

"Unfortunately that's the one thing I haven't been able to grow out of." Bennett fanned his face with his hand. "*Sooo*, have you seen Patrick yet?" And, how could she have caught the blush through his darn sunburn?

"Yep. I'm meeting with him sometime soon."

"Oh?"

"I suggested the three of us get together, but he wanted to meet just with me."

"Yeah, that's for the better."

Hoosier gave Bennett a playful smirk. "I don't know why you guys haven't made up, but I'm gonna fix it. The Three '*Counceleers*' will reign again!"

"You're truly an optimist."

"Incurable…"

He wondered how closely he resembled his old man. Was he destined to become the same loser, the type of man that would leave his wife and child for a receptive prostitute? After all, he must be a loser if his old man left him behind without thinking twice, or, more than likely, without thinking once. Only in the company of Paul did he feel a sense of self-worth, because, hell, anybody was better off than Paul and his three basic life groups.

The life of a loser—well, all be damn if he wasn't going to live it to the fullest, to go for the gusto, as the beer commercial always said. Patrick jumped when he heard Karl's voice.

"It's for you," Karl yelled out, pointing to Patrick, who was standing behind the bar serving drinks.

Patrick excused himself and wiped his hands on his apron. Karl handed him the phone and left the room, shutting the door behind him.

"I'll pay you double," the voice of Ahab spilled through the phone.

"You already agreed to that and it's not enough. I don't think any amount is enough. Guys are shooting at me, the cops are asking me questions. Forget it—"

"I thought you had a plan to detour the cops."

"I do, but what about all the lunatics out there trying to get the reward."

"Look, I need two more."

"You said one more last time we talked."

"I know, but something special has happened. I'll need another."

"Forget it."

"I'll pay you triple…"

Patrick took a deep breath. The shanty he lived in came to mind—a third story, one bedroom, with rust carpet from 1973, lousy plumbing, cockroaches. And then there was his bed, a glorified cot; his car made an AMC Pacer look good; Candice bitching daily about his lack of income from the bar…"Okay, triple, and you leave me alone after two more. I break clean from this crap. Deal?" Candice would shut up after two more…

"That's a deal, Stubb. How soon?"

…He could get something a step above an AMC Pacer…"A little pushy aren't we." The comment was greeted by silence. "The first one this week sometime. Is that soon enough?"

"I'd say that's just about perfect. I wish you greasy luck."

…A neutral carpet, less than two inches tall. "What's that?"

A hearty laugh broke out at the other end of the line. "Stubb, good luck."

Chapter VI

They walked along William Street in a steady stream beneath Bennett's office window at Pequod. People from all facets of life, he guessed by their multitudinous attire—toting a train of signs slapped together with parts readily available at home: poster board on broomsticks, construction paper on two-by-fours, handheld cardboard with jagged edges, obviously torn from a box. It was an angry stream of rapids—white posters cresting the human line stretching clear to the town hall. The voice of the stream—a short staccato chant—rose high and low as waves of individuals walked down the sidewalk directly below the window. The wording on the signs varied, but the message was clear: capture the lunatic preying on the city of New Bedford. Bennett vacantly turned from the window as the last passed by.

He moved to the front of his PC, staring at the network architecture diagram for the next generation infrastructure Pequod would soon require when they moved to a DS3 pipe to the Internet. His concentration had been focused on WAN access switches, should they go with big and extremely dense boxes or multiple lower density boxes for redundancy? He had been working on the details for hours, but once the marchers appeared outside his window, he found it difficult to concentrate. They meant business—he had never observed such an exhibition in New Bedford.

He clicked his desktop e-mail icon and watched as his new messages were retrieved. He grew a broad smile when he saw Tanya's name appear in the *From* column. She had sent e-mail. She was catching on.

Bennett,

Thanks so much for setting up my PC! You're a big help to me and I want to let you know how very much I appreciate it. Can you join me for supper this evening at my home? It won't be much, but it's the least I can do. Reply and let me know.

Sincerely,
Tanya

Sincerely? Was sincerely a term one would use for a friend, or could it also be used for somebody considered to be more than just a friend? Bennett wasn't sure, but as he caught himself thinking about it, he realized it was at best wishful thinking on his part. Sincerely meant sincerely and nothing more or nothing less. His skin tingled as he felt his body in the grips of euphoria and he found he could not rid himself of a Cheshire smile—his face locked for all to see. He quickly replied back that he would be delighted to join her.

Hoosier popped into his mind, he could still picture her trying-to-hide-the-disappointment expression when he went into further detail about Tanya. She smiled and said she was thrilled that Bennett had found someone. The thought caused him discomfort so he moved on. The next e-mail to capture his attention was from Griggs.

Bennett,

Thanks for your reply. I believe you may be off target on your theory. I am very much a *newbie* when it comes to Unix. I'm sorry, but I can't really be of much help. I posted your e-mail response to me in the newsgroup in hopes others may be able to jump in on this mystery. I've attached a copy of the whale file I encountered on my system.

Sincerely,

Michael Griggs "Griggs"
Moon Enterprises, San Francisco, CA

Bennett didn't find what he read very appealing. He didn't like the fact that Griggs had posted his e-mail in a public newsgroup. It bothered his ego to know the *whale* creator would read Griggs' e-mail and know he was giving Bennett a struggle. The other thing he didn't like was that Griggs was of no help in providing any additional information that might begin shedding even a little light on the mystery. But Tanya's e-mail overshadowed whatever bad karma the note might have conjured, so it all drifted away. He copied the attached *whale* file to a floppy that already contained his other two *whale* files but named it *whale.sf* for San Francisco. He then ran a diff program against it to get an indication if it matched any of the other two. He found it no surprise that the third was different from the others. But the size of the file was close to the size of the other two. When he examined its internal content, he found some matching information. The file Griggs had sent could very easily be a member of the set.

"See them protestors?"

Bennett looked up from his PC to see Collin standing in the doorway.

"Yep. Haven't seen anything like it since Coney Island had two for ones on their dogs."

"Well, Boston had its strangler so I guess New Bedford is due."

"Won't be long before we have Dan Rather live at the pier."

Collin went to the window. "That Alfred guy stopped by my house last night. Do you believe it? I wonder if he's gay."

"The old guy that showed up at our meeting?"

"Yep. The old fart smelled like he crawled from a tube of Ben Gay." Collin laughed. "When I woke up this morning my sinuses were clear and I think it was because my house smelled like Vick's vapor rub."

"Funny, he stopped by my house the other night. Must be a lonely old guy."

"His wife sure plays a lot of bridge."

"I kind of feel sorry for him."

Collin squinted as he looked at Bennett. "You're glowing."

"What?"

"You're glowing. What's with the permanent smile?"

"I'm having dinner with Tanya tonight—"

"That explains it. Ain't that the berries."

"It's no big deal." He shuffled the architecture diagrams on his desk. "She's just returning a favor."

Collin smiled. "Yeah, right. What are you having? Any Cajun?"

"I don't really care."

"But the food's half the fun."

"And what planet might you be from?"

Surely the table would be set with the polished silver, and the china he had seen in the china cabinet on his previous visit. He envisioned crystal glasses with images of flames from tall thin white candles, slow dancing inside. A white table cloth, the smell of pasta drifting into the room with steam rising from the colander in the sink, and music, soft, delicate tones painting the room in audio delights. Bennett could sense it all even prior to arriving at Tanya's home.

"I'm ordering in pizza, if that's okay," Tanya said, closing the door behind Bennett when he arrived.

"I love pizza, that'd be great," Bennett said, smiling, the crystalware, silver, pasta, and china, scurrying for a place to hide in his embarrassed mind.

"Follow me, I want to show you what I've done," she said, smiled like a school child holding up a crayon drawing.

Bennett followed her up to the study. Her PC was on and she pointed at the screen and took a seat in the chair in front of it. "I created a spreadsheet to store a complete inventory of everything I own."

Bennett moved close to better read the screen. "Great job, but you don't own much."

Tanya glanced at the three items in the spreadsheet. "Oh don't be so pissy. It's not a bad start, huh?"

Pissy? He'd never been called pissy before. He thought how beautiful she looked even from behind—her fine auburn hairs flowing down to the middle of her back and curling there freshly. Her soft smooth hand, her delicate pink fingernails, clutched about the mouse. And he could see her reflection in the window, her face simplified in the glass, yet losing nothing in its beauty. He smelled her perfume flowing from her neck as he had the first night he met her. But this was different, no freesia scent. He wanted to breathe her in. To have her inside, becoming a part of him. "I like your perfume."

Tanya laughed, and turned to face Bennett. "Not bad, is it?"

"It's intoxicating."

"It's just plain old Chanel No. 5."

"It's nice—"

"Have you heard of it? Is it one of your favorites?"

Bennett smiled. "I'm not much into perfumes, but I do like it."

"Good. I'll have to wear it more often." Tanya turned around to face the PC. "So have I done good here?"

He thought how nice it would be to sit next to her, to cuddle and kiss.

"…Bennett?"

Bennett snapped from his reverie. "Huh?"

Tanya shook her head. "Helll-ohh."

The doorbell rang. Tanya excused herself. He heard her speaking with someone at the door before she yelled out that dinner was served. He joined her in the dining room downstairs.

As the slices of pepperoni pie shrunk from eight to one, Bennett found himself opening up and telling Tanya all about himself. She listened with interest, asking questions only as a story wound down, invariably winding up another.

"So what about high school?" Tanya asked, after Bennett told her about hacking the school district's mainframe in junior high.

Bennett hesitated. Not a pretty time.

Tanya noted the pause. "Or not…"

"Nah, its no big deal. The truth is, I don't recall too much of the early days of high school."

"What? Were you a binge drinker or something?" Tanya asked, teasingly.

"More like 'or something'," Bennett took a sip of ice water and leaned back in his chair.

Tanya offered an exit question. "Would you like some coffee or something?"

Bennett didn't take it. "No thank you." He took another sip of water. "Anyway, to get back to high school, the only things I do remember, prior to the first semester of my sophomore year, are what people have told me happened. My saving grace was that people I knew well, I remembered, but the events were different. I don't remember them. And it was only the two years prior that I really struggled remembering. Every once in awhile, for no particular reason, I recall a memory."

Tanya shifted in her chair to bring her crossed legs out from under the table and move them to the side. She leaned her head on her fist. "What happened?"

Bennett eyed her svelte legs. "Well, nobody is really sure. Apparently I attended a Halloween Party at Wilbur's Point—"

"Where's that?"

"End of Sconticut Road in Fairhaven."

"The road you take to get to the Town Beach on West Island?"

"Yes. Anyway, I disappeared from the party, I was there with my cousin, Patrick—"

"Patrick?"

Yeah, the man with an ego the size of Nantucket. "Patrick Arnold, I'll tell you more about him later." Much later. "Anyway, the police find me at 4:00 a.m. lying by the side of Sconticut Road. Apparently I have a fairly severe head wound, lost lots of blood and so on, and I'm unconscious. They get me rushed to the hospital and patch my head up. I don't regain consciousness until noon or so that day. I don't recall a thing about what happened to me. And worse yet, I don't recall much of what happened to me in the two years previous." But he could distinctly recall that first day in the hospital. Wondering where he was and why he was there. His mother seated in the chair next to his bed with a concerned expression, the sterile, medicine-smelling air in the room, things beeping, voices trailing out in the hallway, the flimsy curtain by his bed, and the parade of personnel poking and prying his body.

"Amnesia?"

"To this day."

"Didn't somebody see you leave the party?"

"They interviewed almost the entire student body, nobody saw me leave."

"Even Patrick?"

"He told me he looked for me. Finally he gave up, thinking I'd gotten a ride home with someone else."

Tanya scrunched her face. "That's very strange. Why would someone want to hurt you?"

Bennett grabbed the last slice of pizza and took a bite. "I don't know. Nobody could figure that one out."

"So you don't recall anything about your first year of high school?"

"Most of it, no. But between my mother and Patrick, it has been pieced back together for me. At least the important things."

Tanya shook her head. "Good thing."

"And that's not all." Bennett paused to see if there was interest in Tanya's eyes, which there was. "I now suffer from epilepsy as a result of the head trauma I suffered. When I have a seizure, I have what is known as a complex partial seizure, which means I experience a change in consciousness during the episode. I enter a trance like state and won't remember what occurred during the episode. It's also known as temporal lobe epilepsy or TLE for short. For some sufferers of TLE, the side effects of an episode can border on psychotic behavior. Van Gogh is thought to have suffered from TLE. The medication I take usually controls the seizures, but the other day I forgot my medication and had a seizure at home."

Tanya looked sad.

Bennett shrugged. "But it's okay. I've learned to live with it. As long as I take my medication I keep it under control. Everybody has bad things happen to them, I'm no exception."

"You've got a healthy attitude about it. I've found a positive outlook is key to overcoming any setbacks. You won't get anywhere in life without a positive attitude."

"For the most part. Sometimes though, in the back of my mind, I wonder what happened that night."

Tanya moved her legs back under the table and leaned in Bennett's direction. "Have you gone under hypnosis?"

"Why would I do that?"

"Maybe your mind knows what really happened, but consciously you are unable to tap into the memory."

Bennett pictured himself stiff in a chair, a gold pocket watch swinging back and forth in front of his eyes, him confessing his every sin. "I don't think that would work."

Tanya reached out and grabbed Bennett's plate. "So you don't believe in that kind of stuff?" She stood up and headed for the kitchen.

Bennett shrugged. "Well, its not that…its…I just don't think it would work."

Tanya rinsed the plates in the sink. Bennett watched her look into the kitchen window, checking her appearance. He thought it looked as though she might be shaking her head.

"So," Bennett spoke up, "what about you? I know so little. We need to talk about you."

Tanya turned from the sink. "Maybe later. Right now I need you to do me a favor."

"Anything," Bennett said without hesitation.

"I created a file this morning on my machine and now I can't locate it. Can you help me?"

"Of course. What was the name?"

Tanya winced, "I don't remember…"

"No problem. What did the file contain?"

"Stuff for my bills. My electric bill is an entry."

"Consider it done."

Bennett jogged upstairs and seated himself in front of Tanya's machine and brought up a search tool. He searched on the keyword *NEPC* for the local power company. File names that contained the word began to appear on the screen.

```
sound.dll
x2drv.sys
bills.xyz
harpoon
```

The last file listed caught Bennett's attention. It looked out of place. He looked over his shoulder and opened it in a text editor and discovered the hieroglyphics of binary data. Buried in the mess he found some strings of text. Enough to tell him the file was native to a Unix operating system, not the operating system running on Tanya's machine. There was no reason for her to have such a file. Her words from the other day stung. He had sworn no harm would come to her from the Internet, that it was safe for her to signup with Pequod. A box of floppy disks sat inches away from the mouse pad. He grabbed one from the back of the box and copied the harpoon file from Tanya's machine on to the floppy and then erased it from her machine.

Darn it. He shoved the floppy into his shirt pocket. An immediate surge of guilt shot through his mind. He had already let her down. Was it his friend who left it? How did this person know about Tanya? The stakes had just risen.

He noticed her browser was open and before thinking, clicked on the history, thinking in the back of his mind, it could provide some indication as to where the harpoon file might have originated. But all he found in the list were sites that related to adoption and foster children. He felt a pang of guilt and quickly released the history.

Bennett's resolve to find out who was behind the charade now intensified, nobody was going to upset Tanya. Nobody. Pissy or not.

"So what do you think?" Paul said, holding his forearm under the flashlight.

Patrick dropped his headphones and moved closer to Paul. Moisture, clinging to Paul's arm from the light fog, glossed the otherwise flat and almost entirely indigo tattoo: a whale, spouting red with a jagged harpoon embedded in its back.

Patrick remained silent as he observed Paul's arm.

Paul laughed and licked his lips. "It's a beauty, don't you think? Designed it myself. Had it put right above my *SEH* tattoo."

Patrick shifted his jaw. "You asked me what I thought?"

Paul's eyes widened. "Yeah…"

"I think you're a moron." Patrick placed his pointing finger against Paul's chest.

"What's your problem?"

"You! If you had brains you'd be dangerous. Who did the tattoo?"

"Razor…"

"Did you ever stop and think Razor might want the reward money?"

"Razor doesn't care about money—"

"How do you know that?"

"He served in 'Nam, man."

"You're so stupid, I'm not even gonna dignify that…" Patrick shook his head. "Don't you think people are gonna see that and wonder why you have it?"

"I'll keep it covered. I'm not stupid."

"Damn, Paul. We're close to getting in some deep crap here—don't you realize that?"

"For what? Killing a few whales? How much time can you get for that?"

"I don't want to do any time! Do you understand?" That and having to live a day longer with the shag carpet in the home.

Paul's eyes dropped to the deck of the boat. He rolled down his sleeve, covering the tattoo. His voice dropped. "Sure, man, I understand," he said sheepishly.

The ocean was quiet. In the darkness, the *Where II* might just as easily have been floating in a pool in someone's backyard. Patrick half expected a Disney inner tube to float by. There was no moon. Stars and large nearby planets smattered the infinite black above.

"It's eerie out here," Patrick said in a quiet voice, now calmed by a sense of pity for his friend. If he got a new car, it would be black…

Paul nodded. He pulled a cigarette from the pack in his shirt pocket. "You believe in omens?" he asked, looking up into the night.

"It's all science…the laws of nature," Patrick replied, himself looking up…Or midnight blue…chrome wheels.

"My grandma used to say that a calm quiet night such as this meant a man was gonna die; that his soul needed the quiet to become drowsy, in preparation for his final resting. To give himself time to ponder the things he'd done in his life in quiet reflection."

"Who's gonna die?"

"I suppose me."

"Why you?"

"A man such as myself is a mistake in this world." The embers on Paul's cigarette burned brightly as he took a long drag.

The thought of the new car diffused in the smoke. "You're just different. The world needs all types."

Paul shook his head and spit off the side of the boat. "You ever see that film where that guy is stuck in a Turkish prison?"

"The one where the guy gets busted for smuggling drugs and sentenced to life and then goes nuts?"

"Yeah, that one. There's a scene in the film where this guy explains to him that he's a bad machine that comes from a factory, he doesn't know he's a bad machine, but the people at the factory know—"

"So what are you saying? That you're a bad machine?"

"Yeah, and you know it." Paul looked directly at Patrick.

Patrick felt a sense of nakedness. "Don't be ridiculous."

"Say what you will, I've figured all this crap out."

"There's a flaw in your thinking."

"I don't think so."

"How come you know you're a bad machine? According to your theory you shouldn't know."

Paul cocked his head.

"*Shhh!*" Patrick held his finger to his lips. He then pointed starboard.

The sounds of a very loud exhale echoed across the water. Others then joined the solitary breath.

"We've got a herd," Patrick whispered.

Paul made his way to the deck and lifted the planks concealing his harpoons. He extracted one and balanced it in the palm of his hand. He took a deep breath and started making his way to the bow.

"Wait…" Patrick whispered as loudly as he dared.

"What?" Paul whispered back.

"How are we gonna get closer?"

Paul shrugged. "Use the engines."

"We'll frighten them off."

"It hasn't in the past."

"It's never been this calm. Turn on the spotlight."

Paul lowered the harpoon and allowed it to hang loosely at his side. He reached the spotlight and flicked it on. He slowly panned the surrounding seas.

"Go back!" Patrick said in a low voice.

Paul panned back left and stopped just as a blow spout shot skyward, the resulting mist almost reaching the boat.

"They're right next to us!" Paul whispered. He left the spotlight and returned the harpoon to a position above his shoulder. With care he stepped to the bow pulpit.

Patrick watched Paul in silence as he stood like a granite statue, arm held high, harpoon readied.

Waiting.

Waiting.

The arm flew forward, thrusting the harpoon downward. The rope by Paul's foot scorched the deck as it ripped across the bow. Paul grabbed the bow railing, Patrick the wheel of his vessel.

"Hold tight, we're going for a Nantucket Sleigh Ride!" Paul yelled out.

The *Where II* silently sprang forward with tremendous force. Paul lay stretched out on the deck of the bow, but clinging to the railing. Patrick felt his head snap back. "Damn!" he yelled in pain.

With little noise, the boat slid across the glassy seas, the whale just ahead, death in tow. When the boat slowed, Patrick watched as Paul struck the whale's exposed hide, resting starboard, with his lance, driving deep for the lungs, hoping to flood them with blood. It was as if Paul were operating in slow motion, driving the lance ever deeper, and Patrick had difficulty in believing he plunged it into some living organic object. A misty fountain of red fell upon the boat as life fled the whale through its blowhole—the whale hopelessly attempting to expel the blood now gushing into its lungs. Paul held his hands up skyward as if inviting the red rain in some gruesome act of redemption. Patrick wiped it quickly from his face, and brought his shirt up over his head. In the periphery of the spotlight, he watched Paul's body drip in crimson, the fine mist of blood spray-painting his clothes and skin. Paul absently finger-painted on his forearm, moving his index finger to form various patterns in the blood on his skin. And he hummed a tune, rocking on his feet, standing at the bow.

A burst of nausea erupted through Patrick's body. He threw himself to the side of the boat and watched his vomit explode into the sea. He gazed in surrealism as his pale vomit mixed with the crimson seas and his distorted face reflected in a mirror of pink—stringy vomit dripping from his lower lip and chin. At his last convulsion, when looking beyond his own reflection, aided by the spotlight, into the sea at a shallow depth, he noted a gray oblong shape, appearing to nestle the dying whale lying starboard. He could see a weak black eye buttoned in the gray shape; he felt the realization in his gut first, and it took a few moments for his brain to catch-up, much like the time he held his first child right after her birth—the gray mass and the frightened black eye belonged to the calf of the slain whale. It remained by her as she died. Nudging.

As he brushed a thick dapple of vomit from his chin and flung it into the sea, he spit and whispered, "You're right Paul," spit again, "*we* are bad machines."

"I'll be right with you," Mrs. Bidwell's voice came from around the corner.

DeWald eyed the floral pastel material making up the sofa he found himself seated on. The lawn mower started up just outside the window behind him. He looked over his shoulder through the white crochet curtains to see the puzzled shape of Mr. Bidwell's obese figure.

"First time in two month's he decides to cut the grass," Mrs. Bidwell's said, still from back behind the corner.

DeWald turned to face the room. His eye caught the glass animal figurines prowling across the top of a broken stereo cabinet with one door—an ancient TV behind it—held by a single hinge. A lone photograph, encased with dust, rested on the dusted surface of an end table next to the sofa. The Bidwells stood at either end of the photo, with three children of various shapes, colors, and sizes, crowded in between. The lamp shaped female in the middle was Sheryl, he guessed.

DeWald pulled his notebook from his pants pocket. Thomas had checked for known predators hiding in the homes along the route Cynthia Mendosa jogged and had come up empty handed. DeWald had called in to the Boston field office to get a computer forensics resource. They said he'd have someone this week. The image of the bearded fisherman shot and killed accidentally, during a hunt for the whale killers, popped into his mind. He had warned them, but the whole thing had also been his idea—that caused him to wince. Never had any case contained so many opportunities for major failures, ones that could cause him to doubt his ability to handle such a case. Was he in over his head?

DeWald took a deep breath. Sheryl Bidwell, age 23, barmaid at the *Beer Trap* and found dead in the rotting corpse of a 48-foot Humpback. Sick bastard. According to Thomas' investigation she was a high school drop out, loose with men, a whore, no raving beauty—DeWald eyed the photo and agreed. Nothing in common with Cynthia Mendosa other then both were female. Whoever he was, his selection of victims was not based on physical attributes or personality, or any other trait of precision. If it breathed and blossomed with estrogen, he'd kill it—

"Harry are you coming in to talk with this gentleman!" Mrs. Bidwell screamed through the front screen door causing DeWald to jump. He looked out the window to see the balloon of a man wave his hand at the door and continue his plodding with the mower.

"I'm terribly sorry. I think he's just avoiding us," Mrs. Bidwell said, seating herself in a rocker across from DeWald. "She was adopted you know…Harry—and I didn't know this before we were married; in those days you weren't supposed to—had very weak sperm. He had the virility of a cap gun—"

DeWald feigned coughing to cover his laughter. "Who's she?"

Mrs. Bidwell pulled her head back and scrunched her face. "Why Sheryl, of course. Anyway, so there's no genetic material in her, you know, from Harry or I."

DeWald nodded. You still raised her. "So what were her hobbies?"

"Let's see, where should I start…oh yes, men."

"Any others?"

"No, just men."

"Did she hang out anywhere other than her place of employment?"

"You mean that whore house? No."

Hostile witness and the mother no less. "So that's how she spent her entire day, everyday?"

"Pretty much. She worked from noon to the early hours of the morning. I thought she was whoring the whole time, but I found out yesterday she was doing other business for the owner of the bar."

"What was that?"

"It seems Mr. Patton...I think that's his name...started a Web business. Isn't that what you call that new thing they're doing with computers these days?"

"I believe so..."

"Anyway—" Mrs. Bidwell looked up a the ceiling, "Is that a spider?"

DeWald looked up and saw a black spot. "I don't think so."

"Okay, but I'm gonna keep an eye on it. Anyway, he was having his girls, including Sheryl, do these obscene—you know showing their all—live videos and sending them over the Web. That explains the hairs in the tub."

"Help me out there."

"She shaved her—"

DeWald held up his hand. "That's okay. I understand now."

"It moved."

"What?"

"The spot it moved, it has to be a spider. Excuse me." Mrs. Bidwell got up from the rocker and disappeared around the corner. She returned with broom in hand and thrashed the bristles against the ceiling. It didn't budge.

"I think it's just a black spot," DeWald said with every ounce of patience he could muster.

"I'll be damned. I think you're right." Mrs. Bidwell continued ogling the spot, her eyes squinting in varying degrees.

DeWald waved his hand in front of her face to get her attention, to no avail. "Other than showing her all, did she communicate with her viewers?"

"It's disgusting."

"The spot?"

"No, what she did. According to Agnes—she's the one who told me about it—the men would type in things they wanted her to do and she'd do them."

"So she did communicate with them?"

"Yeah, I guess you'd call it body language."

107

"But did she send them a response?"

Mrs. Bidwell looked down at DeWald. "Yes, disgusting things."

"Do you know where Mr. Patton operated from, what building?"

"The bar, upstairs. He apparently has a studio."

Unlike with Mrs. Mendosa, DeWald did not feel the least bit hesitant to ask the next question. "You knew Sheryl took drugs?"

"Of course, goes with the territory."

"Can you give me any names of any individuals from whom she might have purchased her drugs?"

She shook her head. "I haven't a clue. She at least had the decency to not bring her trash home with her."

DeWald jotted down some notes. Mrs. Bidwell looked beyond him through the window.

"If he'd had sperm worth a wooden nickel…"

"Do me a favor, get Alfred away from me. If I smell camphor I'll puke," Collin said, cornering Bennett in the back of the research library at the Whaling Museum. The NBWHS meeting had just adjourned and Collin wasted no time hustling from his front row seat to the back where Bennett had stealthily entered, late.

Bennett found himself eyeing Tanya, who stood over in the corner, a cup of soda in her hand, talking excitedly to Adam. Bennett had watched the conversation begin just after the meeting adjourned, Bennett and Tanya were seated together and Adam introduced himself as the two stood to stretch. Adam was known as a ladies man, and so Bennett watched every movement between the two as he vacantly responded to Collin. "Why?"

"The guy's driving me nuts. He keeps asking me to do stuff. Damn, the man's sixty years old, what am I gonna do with him?"

Adam said something that made Tanya laugh. Bennett spoke without looking at Collin. "I think that's a slight exaggeration—he wants to be your bud—"

"Yeah, mine and everyone else's. What's with the guy? His wife must play bridge every night."

Finally Adam left Tanya's side. Bennett waved as she turned to face the back. She smiled and began walking in his direction. Bennett caught a glimpse of Alfred, now standing next to John and moving his jaw voraciously in charged conversation. John's head nodded as if in perpetual motion, as if it would not cease long after Alfred had spoken his last word. He watched as John pointed in his direction. Alfred cupped his hands and said, "Collin, can quote Melville?"

Collin looked at Bennett. "Oh man, did I hear Alfred call out my name? I'm gonna run like a striped ape."

Bennett nodded. Tanya stood next to him. "I think he wants you to quote Melville."

"Say something from Melville," Alfred shouted again. By now the entire room had turned to face Collin.

Tanya spoke up. "You can quote Melville?"

Collin smiled, smugly. "Oh, I know a little."

Bennett tapped Collin on the arm. "Say something real quick. He's not going to give up."

Alfred approached and placed his hand on Collin's shoulder. Collin jerked away. Bennett could see a questioning look in Alfred's eyes and spoke up, "He'll tell you it's an injury from a skiing accident at Breckenridge and that it still bears pain, but his close friends think he slipped in the tub."

Collin shook his head. "Don't go there," he said, and then under his breath, "Plebeians..." He got up on a chair. "Y'all gather round," he said, with his widespread elephantine arms and pulling them towards his chest. He cleared his throat and spoke. "Take off thine eye! More tolerable

Than fiends' glarings is a doltish stare!
So, so; thou reddenest and palest;
My heat has melted thee to anger-glow.
But look ye, Starbuck, what is said in heat,
That thing unsays itself. These are men
From whom warm words are small indignity.
I meant not to incense thee. Let it go.
Look! See yonder Turkish cheeks of spotted tawn—
Living, breathing pictures painted by the sun.
The Pagan leopards—the unrecking and
Unworshipping things that live; and seek, and give
No reasons for the torrid life they feel!
The crew, man, the crew! Are they not one and all
With Ahab, in this matter of the whale?"
Collin eyed Bennett...
"See Bennett! He laughs!"
...and then, Alfred.
"See yonder Alfred! He snorts to think of it.
Stand up amid the general hurricane,
Thy one tost sapling cannot,
Starbuck!" A bow.

The room erupted in applause. Collin took a wobbly step down from the chair.

Alfred cried out encore and raised his voice above the clapping. "That man could sure write a fancy story. Makes me want to grab a harpoon and head out to sea. What about you fellows?"

"So you've been harpooning those whales, Alfred?" Adam spoke up.

"Every darn one of them!" Alfred shouted back. "That is, Bennett and I—"

"Don't drag me into this," Bennett said, waving his hand in dismissal and laughing.

Bennett stepped over the low fence made of two-inch thick chain, encrusted by a virulent rust; the chain, once used as anchoring line for the larger trawlers, defying fathoms of pressure and brine, now sectioned between cement square pillars. At the other side, clover spread throughout the grass, nearly overtaking it. The sun rested on the horizon of the bay, blossoming like a peace rose betwixt slivers of gray clouds. The breath from nearby seas wetted the grasses causing a dull squeaking against Bennett's tennis shoes as he headed to the rock where the ancient cannons of Fort Phoenix were mounted on timbers. They loomed on the hill as large black barrels, molting their thick coats of paint, pointing out to sea. He walked by a bikini-clad woman attempting to eke out ultraviolet waves from the rapidly withdrawing rays of the sun. As he clambered to the top of the rock, he found a Japanese child mounted atop one of the monstrous cannons, his parents snapping photos and begging him to smile, his father standing on one leg and hopping around with a funny face. Bennett focused downward and bent down aside the monstrous cannon pointing at New Bedford proper and studied its belly, looking beyond the imprint of the king's Cypher, for the markings, Patrick had knifed into its thick painted hide many years ago. But the cannon had long since shed the hide and been provided a new one by concerned volunteers some early Saturday morning.

He felt silly meeting Elander here. A mere conversation with the man was uncomfortable enough, but to partake in a clandestine rendezvous could only be considered absurd. He had found a voice message from Elander on his home answering machine. Elander could have walked into his office earlier in the day and mentioned it—nothing short of puzzling. Bennett looked at his watch. Elander was late. He began walking through the cannons. As he leaned over the outer wall and looked below he spotted a man sitting in a yoga position at the base of the hill opposite the side he had come from. It was Elander.

As he drew near, the yoga took a deep breath.

"Smell that air? It's life, and there's nothing better than breathing in fresh life. You agree, Bennett?" Elander's eyes remained shut.

Bennett picked up a pebble by his foot, took a seat in the soggy clovers, and tossed the pebble into the surf. "Yeah, I suppose so..."

"There's so much taking place around us—do you realize that? Life is a voracious entity; inexorably devouring time, perhaps recklessly in its greed, like a flame the wick in a candle, only to die once the wax of the candle is consumed. And once it is, time is gone. The wick length of time for each of us is fixed." Elander's eyes popped open.

Bennett shrugged. "So what are you saying?"

"I'm saying you need to pay closer attention to the details of life. In your insatiable appetite to proceed with life you are missing dimensions that will harm you. Beware, Bennett."

Bennett wondered why he was wasting his time with this. He forced himself to ask the obvious. "Beware of what?"

"You are naively walking through your days oblivious to the danger mounting against you further down the path."

"What are you talking about?"

"Some are not what they seem."

"Such as..."

Elander slowly shook his head. "Pay attention and you'll know."

Bennett felt the wetness from the clovers making its way through his shorts and sprung up. "Whatever you're smoking has short circuited the old CPU."

"If you say so. But humor me a moment longer and answer me one question: What kills a whale?"

Bennett shrugged. "I don't know...old age, giant squid?"

Elander smiled. "I should have asked you how does a man kill a whale. If you harpoon a whale, you might be greatly surprised at what you find inside."

"It can't be!" DeWald slammed the phone down and tossed the covers to the floor and sat up on the side of his bed. The red display on the radio read 5:12 am. According to Laykin, a man and wife team on an early morning outing wrestling up quahogs discovered a beached whale bleeding from deep wounds in its body. Laykin and Thomas were on their way to the scene. DeWald prayed the whale was empty. An odd, unimaginable prayer—beyond the reach of common experience, limited to the fleeting intentions of Jonah's conspirators. He collected his clothing from the chair and dressed as quickly as his foggy mind would permit. A ground-laden cup of coffee from the Hotel's breakfast nook and he found himself behind the wheel of the rental.

He had vowed no more, and now in the silence of the early hours of the day as he passed beneath the streetlights creating an alarming strobe through his windshield, he had to admit there was a very good possibility he had failed his vow. Taking it in the first place was naïve. How was he to put a stop to something he had made virtually no progress in figuring out? What was he thinking? That he could simply wish away another murder? He was no action hero; the case could not be solved in an hour-long episode. It required dogged determination. "It can't be!" DeWald slapped the steering wheel.

Off Route 6 he turned south to Knollmere. The cottages lining the road were hidden in darkness with no signs of life, except for an elderly woman in her robe collecting the paper from the end of her drive. They had no idea that just down the road the huge distorted corpse of a humpback whale was flattened against the soggy soil at South Shore Marshes, and more than likely, inside, the rigid shell of a young woman.

He pulled to the end of the road and caught sight of Laykin's cruiser. He spotted Thomas attempting to uncoil yellow crime scene tape. Mud was splattered on the legs of his white pants. The morning breeze easily lifted the tape from the pliant eelgrass.

DeWald sighed. There would not be a need for a crime scene unless a crime had been committed. They had found a body.

"We've sent an officer to get some stakes," Thomas said, eyeing the tape flapping in the breeze.

"So we have a body?" DeWald asked.

Thomas shook his head. "No."

DeWald held his hand up to his ear. "Did you say no?"

"Yeah."

DeWald took a deep breath. "Yes!" Thank God for answered prayer. "But why the tape then?"

"We still have a dead whale. There could be some damn good evidence. But I guess we could have a copycat."

"Same ordeal, tied by tail and all?"

"Yep."

"Tongue missing?"

"She's a mute."

"Probably not a copycat then, we never released those details. Another reason not to go public."

"DeWald!"

DeWald looked beyond Thomas to see who had yelled out his name. Laykin appeared over the eelgrass, his arm motioning DeWald to join him. DeWald stepped through the mushy soil stampeding a horde of fiddler crabs and releasing the creepy sound of hundreds of crustacean legs clicking together. When he reached Laykin, he found the disfigured flattened corpse of the whale resting like a muddy pond in the eelgrass.

"She's fresh," Laykin said.

DeWald sniffed the air. "I guess so, this is the first one I've been able to breathe around. Still stinks."

DeWald gently pressed his finger against one of the holes in the whale's hide. Mountainous tears of blubber flapped over the cavity of the wound. "Obviously I'm no expert, but someone really kept driving the weapon they were using repeatedly into this wound."

"Whoever it was, they were motivated."

Laykin took out a cigarette and shoved it in his mouth. A heavy gray bristly beard covered his jaw. "Why no victim?"

"The couple must have been around. Scared them off..." DeWald went silent. He scrunched his face.

"What are you thinking?"

DeWald sighed heavily and shook his head. There probably was a body; they just never had the opportunity to jam it in the whale.

"So there probably is a body." Laykin caught on too.

"Unless our killer waits to kill until after he has his whale coffin..."

"Not likely. We can only hope."

DeWald shook his head. "Let's hope." Make that no reprieve, he thought.

The officer arrived with the stakes. Thomas began clumsily pounding them into the soil with a boulder. A light rainfall began.

Laykin labored to pull the cigarette from his mouth. "We do have something this time."

DeWald sniffled. "What?"

"Evidence left behind."

DeWald's face brightened. "What have we got?"

Laykin lifted a large clear bag resting on the ground not far from his foot. He held up the bag.

DeWald peered at the dark object inside. "A boot?"

Laykin nodded. "A boot. Found it lodged underneath the tail. We figure the killer, or one of the killers, got his foot wedged beneath the tail. Something gave him the jitters, maybe the couple, and in his haste to free himself, he pulled his foot free by slipping out of his boot."

"Seems plausible. Can we get any footprints? Maybe show the guy hopped away."

"Not really. This spongy marshland springs right back up. God must've subcontracted the creation of this beach out to freaking Goodyear."

"So this boot might have been there already."

"Doubtful."

"Why?"

"It's dry inside. With the tide coming in the way it does, clear up to there," Laykin pointed back towards the road about ten feet. "If the boot had been here it would have been at least damp inside."

"Very good, Laykin."

"It's going to be a chore tying this boot back to someone."

DeWald looked to his left. A nice beach aligned the shore not more than 100 yards away. "Ever read Cinderella?"

Laykin thought for a moment and then it clicked. "The hell with that."

Bennett sank into an ergonomically correct desk chair in his office. The same one he fought with everyday to maintain a position whereby he felt comfortable, only to have the chair wiggle out of its settings and catch him slumping like a dope. The red light on his phone flashed indicating he had voice mail, but he didn't want to listen to the messages, and it was probably just vendors trying to sell him something Pequod didn't need. His office wore on him like a tight fitting shirt, aggravating and uncomfortable. It would not be a day filled with productivity. A warm inviting sun effused life outside his window with brilliance, drying the day from a morning drizzle. It didn't help his desire to stay inside and work. The phone rang and he was tempted to ignore it, but near the last ring he picked up.

"Let's play hooky, you know what I'm saying?" a voice whispered at the other end. Bennett right away placed its softness with Tanya.

"Huh?" Bennett said. His chair lurched to the right and only the quick reaction of his hand pressing against his desk prevented him from crashing to the floor. He did maintain his grip on the phone. "Hooky?" He stood up and shoved the chair across his office floor.

"Let's play hooky—you know, go to the beach or something…you need relief from that Internet crap, and the beach is what Dr. Tanya orders."

Bennett looked at his watch. He was working, but at a glacial pace—one of those days where he couldn't get the cognitive processes going. Somehow she must have read his mind. He chuckled. "Hooky, huh? Come to think of it, I do have a slight irritation in my gluteus maximus."

"I'll meet you in front of my office in ten minutes."

Bennett grabbed Collin and told him he wasn't feeling well and winked. Collin nodded and reassured him Dr. Appleton could fix him up. Bennett jumped in his car and sped to Tanya's office, following the directions she had given him over the phone.

"The town beach would have been much closer," Tanya said, as she laid her beach towel in the sand, after careful analysis of the beach topology, at South Marsh Beach.

"I know—I just don't like driving down the Neck unless I absolutely have to," Bennett said, doing the same with his towel. "This is a great place," his eyes looking up and down the beach.

Tanya shook her head. She removed her beach robe, unleashing her toned, red bikini clad body, and lowered herself to her towel, adjusting the bikini to align over tan marks, placing her head on another towel she had rolled-up, and held a paperback up over her face. She squirmed until she found a comfort zone. "What a beautiful day."

Bennett glanced at her tanned skin and tight cleavage, thankful that a similar vision, as he imagined it, crossed his mind when he pulled his baggy swimsuit from his bottom dresser drawer, causing him to leave behind the Speedo he wore while on his high school swim team. He pulled his hands up behind his head and laced his fingers together to form a cup to rest his head in. As he lay, his fingers caressed the back of his head until they encountered the large ridge of a scar spanning over four inches across the lower part of his skull. He gently ran a finger back and forth along the grotesque and unnatural ridge. Mentally and physically scarred, and for what? The perpetrator left his indelible hate, a mass of insensitive skin, clinging to his head like a blood filled leach. He felt a sting on his leg and got up to swat at a horse fly. When he returned to his back he avoided touching the area on his head.

He refused to think about it when at the beach trying to relax. He began to allow his mind to acknowledge the sensations around him. The scent of coconut oil from two adjacent women sunbathers, and a touch of ketchup from the fry stand in the parking lot, wafted across his face in the caress of a cool breeze. The gulls talked from above, awaiting a stray fry to find its way into the sand. A radio, two groups down, played indiscernible music. Muffled discussions began and ended. A child screamed with delight and a dog's bark mingled in. The surf methodically stroked the beach, thundered, and withdrew back into the ocean. He found himself on the brink of dozing, drifting, undulating back, retreating with each wave...

"...You can find a quahog because they squirt at you when you step near them," Bennett said to Patrick, his pout stomach pushing out from his seven-year-old spindly body as he carefully stepped across the cushy damp sand in his thongs.

"That ain't quahogs Benny, that's clams," Patrick replied. He stood much stronger and taller at nine.

"No sir! You've got it backwards," Bennett smarted back. "See there's one there!" Bennett said, watching a stream sprout-up in front of him.

"Dig him up—we'll see who's right!"

Bennett bent his legs and squatted over his feet and began to dig. Moments later his hand hit something hard. He gripped it and yanked it from the sand.

"See! That ain't no quahog that's a clam! Ha, ha!" Patrick said, and began prancing about the sand.

Bennett knew Patrick was right, but he wouldn't allow himself to admit it. "This ain't no clam!"

"Here, give it to me!" Patrick yelled, and ran over to Bennett.

Bennett hesitantly released his grip and handed it over to Patrick.

"I'll prove to you it's a clam."

"How so?"

"Clams have soft shells and quahogs have hard shells…right?"

"Yeah, so?"

Patrick bolted across the sand until he reached the rocky corridor running perpendicular to the beach. He skillfully maneuvered his way atop the rocks in his thongs and stopped once he reached the highest one. "I'll prove it now," he said. He pulled back his arm and thrust it down, releasing the object in his hand, and smashing it across the surface of a rock below him. The contents spilled and trickled down the sides of the rock.

Bennett stood aghast. "Why'd you do that Patrick?" He approached the rock.

"To prove it to you. A quahog wouldn't have busted up like that. If it had been a quahog it would have bounced right off!"

Bennett watched as the gooey contents began baking on the hot rock like an egg in a skillet.

"See Benny, I'm right and you're wrong!"

"I guess so," Bennett said softly.

Patrick continued his adroit journey out onto the levy. "Join me Benny, I've got something to show you."

"What's that?" Bennett asked, as he began his own skillful ascent behind him.

"Somethin' me and Paul did yesterday."

"Did you catch an eel in your minnow trap?"

"We did somethin' even better!"

"I can't wait to see this." Bennett began picking up his pace.

Patrick stopped at a rock near the tip of the levy. "Good, it's still here, Benny."

"All right!" Bennett yelled. "I'm almost there!" Bennett carefully made his way to Patrick until he stood beside him.

Patrick pointed a few feet out into the ocean. "Look."

Bennett squinted and followed Patrick's finger. "I don't see nothin'."

"Look down into the water, near the bottom."

Bennett squinted again and with some strain, saw a shape come into view. It was something long and gray, wallowing in the currents below. "What is that thing?"

"A dolphin."

Bennett's eyes popped wide. "Is it really? Why doesn't he swim off?"

"Cause, Paul and I killed him yesterday," Patrick said proudly.

Bennett continued looking at the shape. "Why'd you do that?"

"We hit him with rocks—that's why he's pinned down there on the bottom—we put a big rock on his tail."

"Are you sure he's dead?"

"Yep."

Bennett returned his glance to the lifeless object pinned to the ocean floor. He watched it yield to the gentle sway of the sea's current, back and forth, entrancing him so that he could not pull away from his stare. "He's dancing in his death," Bennett whispered.

"Yep."

"Why'd you kill him Patrick?"

Patrick squatted on the rock. He picked up a flat stone and skipped it across the water. "For the fun of it…ain't you ever played hunter before?"

Bennett shook his head slowly.

"Well then! Next time you can play!"

Patrick stood up and put his arm around Bennett's shoulder and said, "come on…let's go back and see what Granddad's up to."

The two began to walk back up the levy to the beach. "I'll race you, Benny!" Patrick yelled out when the two were halfway back. Both picked up their pace and scrambled madly over the barnacle covered rocks. Patrick pulled ahead and Bennett worked harder to catch-up. As Bennett stepped atop an unusually craggy rock, his foot slipped on the barnacles and he yelled out as he stumbled down the side of it. Patrick heard the yell and stopped. "Are you okay, Benny?"

"No! My leg, it hurts really bad!" Bennett said, his face cringing.

Patrick began retracing his route back to Bennett. When he arrived, Bennett was gripping his leg. Blood poured beneath his hand.

"Is it bad?" Patrick asked, trying to capture his breath.

Bennett shook his head. "I don't know—I'm too scared to look."

117

Patrick nodded. "How about if I look?"

Bennett thought it over for a minute and said, "Okay, but don't tell me cause I don't want to know!"

Bennett slowly removed his grip and allowed Patrick to look at the wound; Patrick's eyes widened in alarm.

"It's gonna be okay," Patrick said, fighting-off his squeamishness.

Patrick removed his T-shirt and rolled it up. He took the shirt and wrapped it gently around Bennett's leg where the wound bled. He then picked Bennett up in his arms and began carrying him up the rocky levy. Gradually he made his way to the beach and finally to Granddad's home. He put Bennett on the front steps and then ran inside yelling "Granddad!" Bennett looked back to the beach. Gulls were perched and pecking at the rock with the smashed clam. The waves caressed the sand where he had walked minutes earlier—

"—Hello, is anybody home?" Tanya said, gently tapping Bennett on his shoulder.

A spasm shot through Bennett's body as he snapped from his dozing. For a brief moment he forgot where he was and why he was there.

"Are you okay?"

Bennett nodded. "You took me by surprise, I was sound asleep."

"Sorry."

"No problem—I don't want to sleep this beautiful day away." Nor return to some stupid dream from my youth, he thought.

"Want to go for a walk down the beach?"

Bennett surveyed the vast ocean before him. "Sounds good. Sounds real good, let's go."

The two headed across the dry white sand, carefully making their way through nomadic clusters of beach towels, coolers, and beach umbrellas. They began walking parallel to the ocean where the surf broke on the sand. Bennett was looking out into the ocean when the bell from a marker buoy briskly rang off shore. A speedboat zipped by the buoy, turned parallel to the shore, and without slowing, headed in the same direction the two were walking.

"Boy, they're too close to shore to be going that fast," Bennett remarked.

"I think that was a Harbor Patrol boat."

"They must be after somebody."

"Look further down the beach," Tanya said, and pointed.

Bennett placed his hand over his eyes and squinted in the bright sun to look where Tanya pointed. A large group had gathered in a small area. The speedboat slowed when it reached the group. It turned towards the beach, slowed some more, and stopped ten feet from the shore.

"What do you think is going on?" Tanya asked.

"I don't know. Maybe a drowning or something."

"Let's go check it out."

Bennett stopped. "I'm not sure I really want to."

"Where's your spirit of adventure? Come on let's go."

Tanya took a few steps while Bennett remained still.

"Come on! Let's go!"

Bennett shook his head, but begrudgingly agreed.

Tanya picked up the pace and Bennett remained a safe distance behind her. She reached the crowd first. They formed a wall in front of whatever was just beyond them. Tanya turned back towards Bennett and waved her hand toward the crowd urging him to get closer. He cautiously drew next to her.

"Follow me," she whispered.

Between saying 'excuse me' and tapping people on the shoulder, Tanya cut her way through the crowd with Bennett in tow. When she stopped, Bennett noticed a yellow tape, two inches wide, pressed against her belly.

"This is a crime scene," Tanya whispered. "Let's get out of here."

"Why?" Bennett whispered back. He looked over her shoulder. Some 50 yards in front of her, a large, long, dark gray object rested in the eelgrass. Police officers in uniform stood guard while other men in suits studied the object and took photos.

"What is it?" Bennett whispered.

"It's a dead whale, probably beached itself," a voice came from the crowd.

"Why are the cops here?" Bennett asked.

"That's the question everybody here has been asking," someone else spoke up.

The men closest to the object were confining their investigation to one side of the great mass. They bent down on a knee and were looking closely at the bottom of the beast. The men were talking to each other, but their words were lost before they reached the crowd. Bennett watched with great interest as they worked. He looked to the left of the beast, up the shoreline, and noted the flashing lights of a police cruiser. Two individuals emerged through the fence and began making their way towards the beast. Bennett watched as one of the men barked out something to the men closely observing the whale. The man then looked to his right directly at the crowd. He pointed and spoke again to the other men. He turned towards the crowd and in a very deliberate gate began walking at them, yelling something as he went along.

"He's telling us to get out of here," someone in the crowd said.

119

"Let's go, Bennett." Tanya said, quickly moving back into the crowd.

Bennett did not move. The man drew closer. He was within 20 feet of the crowd when others began peeling back. Bennett could now clearly see the man and his navy parka with the bold yellow letters: FBI.

Somehow the afternoon escaped, leaving early evening front and center, complete with the orchestrated methodic chirp of crickets filling the dark night outside, as Bennett pushed open the window in his office at home. The cool damp air felt soothing to his ruddy skin. It had been a relaxing afternoon, but the burn he felt presently made him question his decision to go. He powered up his computer and watched the boot sequence begin. Of course anytime with Tanya was worth it. His feelings for her were growing; she permeated his mind. He found it difficult to form a thought and leave her out of it. On the return trip they had discussed the possibility that the beached whale had been harpooned. The crime scene tape sealed it for them; it must have been killed like the previous whales.

Someone on a motorcycle gunned the engine and shot down the street beneath the window, startling Bennett. The computer finished its boot sequence. He pulled open his desk drawer and retrieved the diskette he had taken from Tanya's house. He popped it into the floppy drive and viewed the contents. He felt a tinge of guilt as the listing contained files other than the *harpoon* file. In his attempt at hurrying things along, he had failed to check if the floppy he had grabbed was blank.

Patton1.jpg
Patton2.jpg
Patton3.jpg
Patton4.jpg
Patton5.jpg
harpoon

He thought they must be account files for Millers, and he fought off the temptation to view them; they were none of his business. He located the earliest *whale* file he had copied to his machine, and moved the *harpoon* file to the same directory, removing it from the disk. Harpoon and whale, perhaps they worked together, at least Elander, wearing his Nostradamus hat, had predicted it so. He took a stab at it.

> harpoon | whale

The command did nothing other then to pause and wait. He performed a break and thought he would try the reverse.

> whale | harpoon

The command had the same results. He did another break and tried a third option.

> harpoon whale

The hard drive on the computer began to churn. The command was doing something. When it stopped, Bennett listed the files on his hard drive.

> whale ambergris

He squinted at the display and took a second look. His jaw dropped as he noted the new file now in the directory. What was it? He knew about ambergris. He had seen products made with it in the whaling museum. But what was the file? It made some sense, ambergris came from inside a whale, a sperm whale, and perhaps if harpooned and killed, one might find ambergris inside, whalers did and treated it as a treasure greater than gold. But why ambergris? Why not a heart or some other well recognized organ?

Bennett opened a hex editor to view the ambergris file. He recognized the signature of a GIF image file in the header portion. He felt his pulse increase as he opened the image in his browser. It was a large file and it displayed slowly.

When it finished, Bennett jumped up from his chair and staggered back, falling to the sofa; his sunburned face turning an implausible white.

"What in world…"

Chapter VII

The orange flame from the burning candle danced his bulky shadow against the far wall. He slid the tiny silver key into the Master lock, turned the key right a quarter turn, pulled the lock loose, grabbed the handle to pull down the fold up staircase to the attic. The ladder creaked as it unfolded to his feet. He grabbed the candleholder, ascending the ladder, it groaning beneath his feet. Thunder rumbled in the distance, subtle, more felt than heard. Inside the attic he withdrew the ladder and pulled the entire mechanism back up until the door lay flush with the floor. As he did so, it caused a slight jarring in the floorboards and sent an item leaning against a nearby wall falling to the floor with a dull tone. He reached out and collected the item, hefting it over his shoulder. In the candlelight the menacing black rod of the harpoon cast an ominous shadow against the unfinished wall, interspersed with crooked nails, whose shadows marched as an army on the peach timber in which they had been impaled. The flame moved little as he crossed the floor, stopping at a dresser with a great mirror, towering into the cobwebs draping from exposed rafters. The handsome polished cherry of the dresser reflected the flame from the candleholder resting on it. The flame stood motionless, erect, in the stale dry air, grayed with dust, barricaded inside for many seasons, and perhaps once exhaled by the original owner, a mercantilist. Two paths well tread—each beaten into the dusty floor like a traversed walkway to a home on a snow-powdered morning—one to the dresser, the other, disappearing into the darkness of a far corner.

Spread atop the dresser with great order in mind, a glass container of absorbent cotton, pliable wax, cake makeup, modeling clay, crepe hair, and a styrofoam bust lighted by a mousy colored wig. Next to them, an aged box, with the lid hanging down one side, marked theatrical makeup. He looked in the mirror watching the orange glow from the candle play off his unimpressive features. The face was inadequate for his needs. He looked in the mirror over his shoulder, like cat's eyes absorbing the headlights of a car, from the dark corner came two tiny dots, blinking in unison, on then off, on then off.

The eyes were of his captured prey.

They tacitly signaled fear, in a code only the mind could comprehend, the mouth gagged and unavailable. He watched in pleasure as fear, in its purest form, consumed the body, masticating in its jaws the psyche, ripped from the vessel quivering in the corner. Locked in the utmost far corners of his mind, in places only tumors prevail, lingered a wretched and insipid drive hungering for refreshment from unspeakable crimes with appetite so insatiable that fantasies, no matter how vivid, were considered predigested and unpalatable. In weeks previous he had come to the room, the dark corner empty, and relived his crimes of conquest, a delicious affair seasoned with derived fits of anger, fantasies of power, and fabricated acts of manipulation. The eyes behind him were there always, whether real or conjured from the past. But the cravings called for a new experience, a new terror that could only be squeezed from a living vessel in the grip of the rarest and strongest fear: the keen awareness of impending death. The eyes were real today.

Light flickered through a small window at an end of the attic, followed by a loud clap of thunder. In the flash he caught her in silhouette, hunched in the chair, her long hair draping over her knees. He responded with a thin smile. He could taste the fear dripping from her forehead. Delicious. He began to apply the makeup, without conscious thought; his hands knew what to do. They would recreate his face to manifest the dark terrible thoughts in the corner of his mind. The eyes behind blinked and he heard a muffled moan—in the split second, a moment of compassion:

"I'm sorry you've been in the corner so long," he said, continuing to apply the makeup. He stopped and turned around. "I don't usually keep you there that long. But you see, I could not leave you in the maw of my captured prize, people were watching. People are such curious creatures. I wish they'd leave me to my pursuits, but I'm afraid they don't like what I'm doing very much. Odd that they would feel that way, given the good I'm bringing."

He swung around to face the mirror and returned to the application of the makeup. The hands moved with deliberation, as in ritual, like the hands of tribesman painting his face before a great hunt. As the old face succumbed to the new, the secrets locked deep in the mind, began to tread unfettered, the terrain much more to their liking. The flesh colored base applied, nestled around the collodian scar, a rip from the corner of the eye to just above the lip. He smiled and bit his lip, a small tear protruded from his eye; the hideous scar moved him.

Once his hands completed the face, he would kill. "But fret not, there is still much to do," he said. He lusted for the banquet behind him, as he had before. The death of the barmaid was nothing short of orgasmic. She was a survivor and he knew it. To remain as long as she had in the profession she was in, one had to be a survivor. Her feisty nature made the control of her all the more spicier. And as he choked the last breath from her gasping mouth, her eyes popping wide and turning to lifeless polished stones, he felt a rush consume his body, flooding his veins with tickling carbonation.

With forethought, he placed dollops of skin glue about his jawbone area. He pulled open a squeaky drawer and withdrew a limp hairy object. He placed it against the glue, and at that instance, sprouted a disheveled salt and pepper beard. From the same drawer, he pulled a flesh colored cloth. He snapped it over his head and became bald. From the white styrofoam bust, he removed the gray colored wig, placing it on his head. He rotated his head slowly from side to side, never removing his gaze from his face in the mirror. He nodded with approval. He removed a bib from his shirt and brushed beneath it with his hand. He scooted his chair back against the wailing floorboards and stood up. He turned his body from side to side, gazing into the mirror. He retained the smile. He reached down and extracted a black book from the bottom drawer. Holding it in the palm of his hand, the book opened, as if guided, to some point in the middle. He eyes, black, and sunken like shallow caves in his forehead, captured the handwritten scribble on the yellowed pages. The head tilted back, only a little, and the eyes fixed on the atmosphere above, until the book clasped shut in his hand and he slid it back to its cage.

The eyes behind blinked in iteration.

From the same drawer where he retrieved the beard, he pulled a thin strand of monofilament fishing line—20-pound test, fluorescent blue, friction and bird-nest resistant, one year limited warranty—and slid it across his hand, as if presenting his choice of weapon for killing, to her for acceptance, like a waiter a bottle of wine to a diner. The smile dropped, and the face, yellowed by the stagnant flame, went blank, no emotion, unreadable. He turned to face her. His chest moved as a piston, his stare stoic. His hand drifted to the top button of his shirt, which he proceeded to unbutton. He did so with the remaining buttons as well, pausing between each white button, as if he were in the process of revealing some great discovery. He allowed the shirt to flutter from his back to the floor, where it lay crumpled behind him. Three manila colored spans of tape pulled something cylindrical tight against his chest. The tape followed the shirt to the floor.

The eyes followed the motion and froze.

The action exposed a silky smooth and near milk white appendage sprouting from his left shoulder near his arm, dangling like a flaccid, well-hung penis, halfway down his chest. It bore a certain innocence, a porcelain purity, as if it had been broke from the figurine of a smiling Hummel boy. And now it hung from his chest.

Without deliberation, he began down the other path, the appendage swinging with his gait, gently wrapping against his chest. He took the line and wrapped it within his clutched fist, the other, in motion to do the same.

"I'm sorry I have nothing to feed you to," he said, moving closer. "But I will have one. You shall become pure, you shall become the ambergris, the sweet scent of femininity, buried in the decadent bowels of the mighty whale."

The line wrapped once around her delicate neck...was wrapped another...and another.

"You are the ambergris, the sweet scent of femininity..."

"Should I call the police?" Bennett asked. He had Collin on the phone and a printed copy of the ambergris image resting in his lap. The woman in it was spectral, almost blue, her eyes shut and arms folded across her chest. A red ring embedded about her neck, and the tattoo on her forehead, brought the only color to her pallid skin. There appeared to be sand in her dark stringy hair. Bennett guessed she was lying on a beach because the flash from the camera taking the image exposed sand in the background behind her upper torso.

"How do you know she's dead?" Collin asked, after some thought.

"She's blue, man. She looks like a ghost," Bennett turned the page over, not able to stomach the image.

"I think our friend has fried one to many okra."

"I think it's his brain that's fried."

"Well, the death stuff is in with cyber gangs out there, you know what I mean. Our friend probably belongs to one. If I were a betting man, I'd say it's some artsy-fartsy photo. Some freaky S and M crap."

"But what if it's someone who's been murdered. Maybe our friend is a killer too."

Bennett heard Collin rearrange the phone. "You're gonna look like a fool if you're wrong. Besides, say this woman really is dead, the file could have been sent from Seattle, or Florida, or even some European country. Do you think the local cops are gonna do anything about that, let alone know anything about it?"

Bennett flipped the image back over and took a deep breath. "So ignore it?"

"Look, do this—you have other files right?"

"Yes."

"Open them up. See if the other images don't make a better case one way or the other."

"Makes sense, but…"

"But what?"

Bennett sighed. "When I used the harpoon file to extract the ambergris file from the whale file, it removed itself—"

"You didn't make a copy?"

Bennett ran his fingers through his hair. "No. I got caught up in what I was doing. I moved it from the floppy rather than copying it. It was stupid, I know—"

"Jeez, Bennett. Where'd you find that harpoon file anyway?"

Bennett quickly contemplated whether he wanted to tell anyone where he had found it. He trusted Collin, *but*…he thought about Tanya, he would have to have Collin swear himself to secrecy. Collin agreed to it.

"I found it on Tanya's machine."

"Tanya's?" His voice was filled with surprise.

"Yeah," Bennett replied, sheepishly. "She had a fear about connecting to the Internet, but I swore to her it would be safe. Then the other day I found this harpoon file. I quickly removed it—"

"As in deleted it from her machine?"

"Yes, it's gone. I have no copies now."

"Can you recover it?"

He sighed. "Maybe, if I can get to her machine." He paused. "I figure our friend knows us more intimately then we might have first believed. Obviously he knows about our acquaintances."

"Could be a coincidence."

Bennett heard Collin chew on what sounded like a chip. "Come on, there's no way—"

"Hey, how do we know that that same file is not on the hard drives of thousands of our subscribers?" Collin interjected, pausing from his chewing.

"It's possible—"

"Damn straight it is."

Bennett heard the whine of a modem in the background.

"What are you doing?"

"Dialing into Pequod, I want to see something."

"…Checking the logs to see if you can spot any activity that might support what you are saying?"

"Something like that…"

Bennett heard Collin's modem connect. He removed the phone from his ear as the crunching of chips blasted through. Once again he flipped the image over to review it. The woman's lips appeared blue. Did women wear blue lipstick? He'd seen dark colors before, but never blue. She sure looked dead, almost like a manikin—

"Interesting. Very interesting." The soft tinny voice of Collin came through the phone.

Bennett replaced the phone to his ear. "What's interesting?"

"How Tanya got that file."

"What do you mean?"

"Just that. According to my logs I only show one very brief login."

"Just one? What's the date?"

Collin gave the date. It was the same date that Bennett had been at her house and the approximate time he had logged her in to check the account was working properly. But how had she sent him email for the dinner invitation and surfed the net? He realized he had not checked the address she sent it from, and as far as surfing went, she could have used another provider. Too late now, he had deleted the email and compacted his inbox. The time and date stamp on the *harpoon* file had been days after he logged in with her account.

"So how did it get on her machine, Bennett?"

Before he could stop himself, Bennett lied. "All it takes is one login."

"Was the time and date stamp for the *harpoon* file in sync?"

Now he could stop himself, but he didn't. "Pretty close."

"Well, that could explain it then."

There was a moment of silence. Bennett found his mind in a daze as he contemplated this latest information. "So you don't think I should call the cops?" He said, ready to bring the conversation to a close.

"Up to you. But I wouldn't."

Bennett said goodbye and hung up the phone. Tanya had fought Bennett on setting up the Internet connection to Pequod, yet she herself had set something up with another provider. What sense did that make? His friend, the whale file dude—how had he known how to get a file to Tanya without the Pequod relationship? Maybe someone physically put it there. Did she have a friend? Could this friend have put it there? What if she put it there…? No he reasoned, not possible. She had nothing to do with it. It had to be that way. There was some other explanation. But what could it be? Whatever it was, no matter how hard he had to think, he would think of it.

He thought of the other files. For some strange reason, he felt the need to review them. After all, he was justified now, there were grounds to be cautious, to protect Tanya. Perhaps they would shed some light. He mulled it over, all the while realizing he was searching for justification.

He found himself in front of his PC, the floppy disk in his dubious hand. He slid it in and listed the files. With the harpoon file gone, only the Patton files remained. They were image files so he brought the first one up in an image viewer. As the image loaded from top to bottom, it unveiled skin and more skin. It was a pornographic image of a blonde-headed woman with her legs spread-eagle. Bennett felt a sickness coming over his stomach. All of a sudden his image of Tanya changed. What would she be doing with pictures like this? He forced himself to look at the others. The woman in the last photo looked familiar, like the woman in the ambergris image. He brought his printout next to the screen and eyed each. They were the same, had to be.

He shut off his machine and walked downstairs. He opened the door to his back yard and flung himself on the chaise lounge where days earlier he had sunburned himself. The night was cloudy and he could hear thunder in the distance. He folded his hands behind his head and stared at what stars he could see between the clouds. A sprinkle started, and the thunder grew louder. He wiped the moisture from his face. The sprinkle turned to a downpour. Wiping his face became a futile motion, so he laid his arms by his sides, the rain now pounding his body.

A newspaper held over his head, he adroitly avoided the puddles forming in the parking lot from the cascades of rain tumbling out of unseen clouds in the night, only to slip in his attempt to stop suddenly at his car, and fall into a small pond that formed in a patch of sunken asphalt. Patrick laughed as he watched the drenched man stand and kick his foot through the water and cuss so loudly Patrick could hear him on the other side of the window glass. He turned his eyes to the door of the Crab House and caught a woman looking nervously over the restaurant. Obviously the man's wife, hoping no one had witnessed the gymnastics. Patrick waited in great anticipation for her eyes to rotate in his direction, and when they met his, he smiled as big as he could. She looked away, out the glass of the front door. Patrick laughed when he heard the loud honk of a car horn and observed the woman frantically trying to open the front door.

"I love the rain," he said, turning his glance to across the booth where Hoosier was seated. "So you found a place yet?"

Hoosier took a sip of her ice tea. "Got something in Fairhaven. Nice place over on Maple; quaint, just what I was looking for."

"Fantastic. Good old Fairhaven. Brings back some fond memories." Patrick ran his fingers up and down his glass of beer. "So when does camp start?"

Hoosier sat up in the booth, her face wreathed in a broad smile. "I'm so excited. We had our first meeting—you know planning out what we're going to do—it went so well. Great staff. The only drawback is Theresa, I don't know if you remember her?" Patrick nodded. Fat broad. "She's still there. At least I left for a decade. And she's still the witch she's always been. And she's huge, must weigh three hundred pounds, loves to throw her weight around, pun intended!"

Both laughed. "So, you miss home yet?" Patrick asked.

"Nope. I'm doing the right thing. I truly believe that."

Patrick leaned on the table and looked into Hoosier's eyes. "I'm glad you're here. I missed you."

Hoosier nodded. "Thank you." She lifted her ice tea in a toast. Patrick clanked her glass with his beer glass. "To the old days," she said.

Patrick gazed out the window. Thank you—that's it—no I-missed-you-too? The rain pummeled the parking lot in sheets made visible under the towering lamps on the parking lot perimeter. There was something about Hoosier that made her a natural when it came to someone who others might gravitate to when seeking an ear for confession. He wanted to spill his guts over his recent activity. The last kill had stuck with him, bringing on alien feelings that caused him some discomfort. He could still clearly see the blood showering Paul and his boat, and from the transparent water at the side of the boat, the calf directing its eye upon him. Something had brought his conscience up a notch to a higher level of sensitivity, but what? Years ago, he had wanted to talk to her about himself when they were counselors, but found he could never be honest with her, and he had thought, going into having dinner with her, that things had changed and he could do so, but he found this not to be the case.

"Huh?" Patrick said, realizing he had been asked a question.

"What's on your mind?" Hoosier repeated. "You look like you're deep in thought."

The waiter stopped at the table, dropping off two piping hot lobster dinners. Patrick hoped the food would be a distraction. He cracked a claw and began extracting the tender white meat from it.

"Something's bothering you," Hoosier said, and took a tiny nibble of lobster.

"Nope, not a thing."

"You've got that constipated look on your face."

"Constipated look?"

"Just like you did when we were much younger."

129

Patrick took a long swig from his beer. He thought about revealing some of the truth, not all, just enough to make his problem known without all the exact details. "Let me tell you a tale," he began and paused as Hoosier began toying with a claw.

"I'm listening," Hoosier said, maintaining her eye contact with the claw.

"A few months ago I'm sitting at my bar minding my own business playing a hand of cribbage with Karl. And then this really weird guy sits next to me—"

"Weird in what way?"

"Weird looking. He almost doesn't look real—I mean his face looks plastic-like, like its severely burned, you know, and he has this long scar," Patrick said as he positioned his finger above his eye and dragged it down to his mouth, "like this."

"Yuck." Hoosier scrunched her face.

"So anyway, this weird looking guy just sits there watching Karl and I play cribbage. I ask if I can help him. He says no, he just enjoys watching a good game of cribbage. I ask if he plays. Again he says no—but he just sits there."

"So the guy won't leave?"

"Right. Finally I pull a lurch on Karl and get up to move behind the bar. He grabs my arm before I can get past him and asks if he can chat with me—"

"Darnit!" Hoosier said, her eyes were fixed on her blouse where a big brown stain was expanding. "I spilled my ice tea."

"Just put some water on it," Patrick said, scooting her ice water closer to her.

"I've got to go to the ladies room." Hoosier got up from the booth and hurried her way around the corner.

He shook his head and grabbed another bite of lobster. He had started the memory of his first encounter with Ahab and now with Hoosier gone, it still played in his mind. The man standing there, rolling a candy in his mouth, as if he required a bit more pleasure from its flavor before he could speak. When finished, in his fruity breath, he asked Patrick if he had a boat. Patrick thought it odd the man failed to introduce himself and would start off with such a deliberate question. He meant business, regardless of his pursuit. Patrick told him he had a boat, but asked why the man wished to know. The pace of sucking slowed as the man seemed to ponder Patrick's question. He asked Patrick if he liked to fish. Patrick shrugged and told the man he guessed so. Patrick recalled the grotesque expression on the man's face. It was as if the man's skin was loose on his skull, like he could grab it and wad it up like a rag. The sucking returned to its initial pace and Patrick was asked if he would like to go fishing. Patrick told him he didn't have time for fishing. He received a chuckle from the man. He offered Patrick one hundred dollars to take him fishing.

Patrick recalled his astonishment at the request. The man could rent a nice boat for that amount. "Why me?" Patrick had asked.

"I have some plans. Plans that will make you some money."

"How's that?"

"Are you interested in making good money and having an adventure at the same time?"

"Good money? Yes. Adventure? You mean like exploring the freakin' Congo?"

"Are you willing to do something mildly illegal for me?"

Patrick looked the man up and down. "As in? I won't smuggle drugs—"

Anger crept in the man's voice. "I'm doing this to rid—," he paused, appearing to collect himself. He popped another candy and rolled it around a time or two. "This is something else. We can discuss it on our fishing trip."

He should have never agreed to the trip, but a couple of days later, he found himself at sea with the man. Patrick could detect the man sparring with him, feeling him out, checking if he were capable. At first the man's idea sounded ludicrous: killing whales.

"For what purpose?" Patrick had asked.

The reply was it was none of his business.

When the man insisted the whales must be harpooned, Patrick started the engines on his boat and headed the boat towards shore. The man tapped his shoulder.

"Where are you going?"

"This is crazy!"

"You've killed Flipper, what's a whale to you?"

Patrick's face went stern and his eyes flashed to the deck of the boat, he throttled down the engines. He cleared his throat. "Just who the hell are you?" he said, returning his eyes to look at the man.

"You're surprised I know that, aren't you? I know a lot of things."

"You're giving me the creeps."

"I'll pay you one thousand dollars for a nice freshly killed whale...a sow, I want a sow."

Patrick shifted into neutral. "One thousand dollars?"

"Yep."

"A thousand bucks...I don't know the first thing about harpooning a whale."

"Go to the Whaling Museum library. All you need to know is there."

"And where do I get a harpoon?"

"Your friend Paul—"

"How do you know Paul?"

"I know many things. But that is not the point; the point is Paul makes a damned good harpoon. I suggest you enlist his help. Split some of the money with him."

Patrick made his way to the tail of his lobster. Hoosier remained in the ladies room. As it turned out, Paul was making harpoons and his exact remark when asked about it was 'awesome!'. Patrick told Paul at the time they were being paid 500 hundred dollars and would split fifty-fifty.

Hoosier appeared from around the corner and trotted back to her seat, a big wet blot on her blouse. "Sorry," she said out of breath. "So what was next?"

"Next?"

"The weird guy at your bar…"

Patrick told her the tale, substituting the whaling with illegally hunting tuna by means of using dynamite. By Hoosier's expression, she bought it.

"So now I don't know what to do. The money's good, but I'm concerned what this guy might do if I say no."

"You mean hurt you some way physically?"

"He's some freaking computer wizard. Who knows what he might do."

Hoosier looked puzzled. "What could he do?"

"From what Bennett's said before, you can do all sorts of things: ruin my credit, take the bar away, next time I'm having an operation have my nads cut off."

"Maybe you should go to the Police—"

"And go to jail?"

"For blowing-up tuna? Since when did that become a felony?"

"Believe it or not, you can do time." Patrick knew he was making something up, but it could be true. His hopes of her making him feel better were evaporating with her encouragement to do what was right.

"I still say go to the Police."

Patrick nodded. "Let me think about it." Ain't no way.

"Well, if I can help, let me know." Hoosier motioned the waiter over for more ice tea.

"Not a word, Hoosier," Patrick warned.

"I won't say anything," Hoosier said.

Patrick gripped her hand that was resting on the table and gave it a squeeze.

"So are you looking forward to the visit to Granddads?" she asked.

Patrick pursed his lips. "Sure. Except Bennett will be there."

Hoosier sighed. "What happened to Bennett's father was not Bennett's fault."

"Everyone keeps telling me that, but answer me this: if Bennett had not gone out in the boat that day, would my uncle be dead today?"

"Maybe yes, maybe no. The point is you can go at anytime. How do you know he wouldn't have died of cancer two years later."

"That's all hypothetical—"

"And your question about Bennett's dad being alive today if Bennett had not gone out on the boat, that's not hypothetical?"

Patrick leaned forward on the table. "Look, the death of my uncle is a direct result of Bennett taking the dinghy out in those choppy waters—period!"

"...This is WPRO. The time is six-thirty. Today's weather calls for sunny skies with highs in the low seventies. Gosh it's good to have some warm weather, isn't it guys? Boy, I'm gonna have to breakout my Hawaiian shirt. Okay, we've got late breaking news from our favorite guy, that Governor guy, what's his name—can somebody call in and give me the guy's name. Seriously though, and I really mean it this time, we've got to get this wacko off the streets of New Bedford and I think Governor Gilmore is going to help us out, and for that I applaud him. So without further ado, the following is a message from the Governor's office..."

Bennett rolled over to the side of his bed to position himself within striking distance of his radio-alarm clock.

"...Frank Gilmore, Governor of Massachusetts, has scheduled a 9:00 am press conference to discuss the latest developments in the recent serial killings taking place in the town of New Bedford. It is believed the Governor will make a strong statement urging the public to get involved in apprehending the murderer or murderers and that he will make a statement indicating that he will do everything in his power to ensure those convicted should receive the death penalty."

The DJ paused. "Now why did I just have a mental image of Governor Dukakis in that army tank? Did we share an image there folks? Now wait a minute...wait a minute, putting on my serious face here because this is a serious matter. Okay, no more joking now. Now I'm sure our listeners are wondering how this can be accomplished considering we have no death penalty in this state. Well, that puzzled me too, but it appears that the Governor is going to request special dispensation from our current laws on the books. I don't know if he can do this or not, but I guess it's kind of like an anti-pardon. I'm not sure I understand it, but if it allows us to fry this guy like the yellow-fin tuna at *Cap'n Bill's House of Fish*—well, I'm all for it…"

The analogy brought home the scent of hot oil to Bennett; he shuddered in the cool morning air. He sat up on the side of his bed and looked in the bathroom at his clothes he was wearing the night before, drying on the shower curtain rod. For thirty minutes he had remained in the chaise lounge, drenched in the downpour. What he found, the Patton images, disturbed him greatly. He realized how little he really knew her. What he didn't know about her, he had filled the gap with his own fantasies on what he wanted her to be, and now those fantasies had been exposed for what they were by the reality of what he discovered. He dragged himself to the kitchen and started a pot of coffee. He looked out the kitchen window, the neighbor's silver tabby, Toby, sat perched on the fence, eyeing a perky cardinal lighted on a power pole, his jaws chattering in excitement. Bennett rapped on the window, but Toby ignored him. How could a cat be so mesmerized by something he could never have?

But he's got high hopes!

The tune about the ant moving the rubber tree plant spun up in his brain causing him to wince. It played like fingernails on a chalkboard. He mentally ripped the needle across the vinyl disc stored in his mind, wishing he could banish it from his memory forever. If only his mind were a computer operating system that he could control from the command line. Heck, he'd even take a crappy DOS command line:

delete HighHopes.mid

Poof!—the tune is gone forever, but why stop there…

delete AntRubberTree.jpg

In the bit bucket—the tune and the image! Doc! I can think again!— well at least he still had his sense of humor.

The coffee maker sputtered out the remaining coffee and Bennett poured a cup. He seated himself at the kitchen table and nursed the mug he had acquired at a gift shop in Woods Hole. He found himself lost in chaotic thoughts going nowhere. He needed a distraction, a pursuit. He thought he should check on his friend, the one sending him the whale files.

He went upstairs, turned on his PC, and headed directly for the newsgroup where he had found the information from Griggs. Someone had posted a new article:

"Hey moronic dolts! I've seen smarter fence posts. I practically give away the keys to my kingdom and you still are in the dark. And you, the idiot at Pequod, you're a has-been! You're so out of it, you don't know how out of it you are. It doesn't seem fair in a way...oh well—maybe someone will help you, perhaps someone mentally handicapped. By the way, if the sun should not rise tomorrow, and you were to catch me—well, let's just say you'd know everything, both for today and tomorrow. But what are the odds? Millions...billions to one? But I'm being too generous, time to sign out.

Mocha Dick"

"We got us a real clown," Bennett chuckled. This guy was something else. Mocha Dick? Well I've got news for you Moby or Mocha, don't call me Ishmael, call me Ahab and your butt is mine. Bennett combed the header of the posted article for any clues. It was of no surprise to him to find all the information in the message header bogus. Yes indeed, this guy was a real piece of work.

Bennett got up from his chair and walked over to the bookcase against the wall perpendicular to his sofa. He pulled a copy of Moby Dick from the shelf and returned to his chair. He clicked on the button to post a reply to the article. He recalled when he had read the book in college that the exciting drama occurred towards the end of the book. He flipped to the back and carefully searched the text. He found what he was looking for and began typing verbatim:

"Towards thee I roll, thou all-destroying but unconquering whale; to the last I grapple with thee; from hell's heart I stab thee; for hate's sake I spit my last breath at thee."

And then added:

But, unlike your ivory-legged foe, my harpoon shall strike true.
You'll find me at Pequod"

Bennett smiled and clicked the send button. He decided to turn his attention to a search engine and see what a search on white whale, Mocha Dick, and Moby Dick might turn up. He watched as the search engine returned with 37,465 potential matches. Combing the first page of matches by reading the attached summaries, he found references to Moby Dick, the book, restaurants, tee-shirt shops, a private insurance company, and other related commercial entities. He smacked his lips and shook his head, looking over the unfathomable number of matches. Where have you gone Mr. Internet? I thought you were doing great when WebCrawler brought back three results. Too much of anything is no good.

The presumed author of the whale file code came to mind. He pulled a black notebook from his desk drawer. The notebook contained his scribbling from things he had noted when reviewing the code. He came across Morye Malchik. The name was entered and the search began and ended summarily with few results this time. The first two summaries were in some foreign language undecipherable to Bennett, but the third held some promise. Morye Malchik was the name of a Russian fishing vessel. The story told how the boat had fished illegally. Perhaps no significance, but the name itself must be Russian. Why would a Russian have any interest in Pequod? Was the woman in the *patton* and *ambergris* files a Russian too? Could everything, like Collin thought, be happening halfway across the globe? How did Tanya get the files? It made no sense.

He thought for a moment to consider his options for searching for more information. Looking around his office reaching for ideas, his eyes passed the bookshelf, where he kept his software; an idea began to flourish when he spotted his code compiler. He could build an automated search agent and set the agent adrift from Pequod out onto the Internet. The design of the agent would be to search for specific information he requested of it and report back to him with any findings. He would decide if the findings were significant enough to follow-up on.

He sighed, thinking about the work involved to build his own private agent. His motivation was running low. Why should he care? Some guy is playing games and in the grand scheme of things what does it matter? He thought about the email. We got us a real clown.

But emotionally he needed a pursuit. He would make this do. The hard drive whined as the code compiler was loaded up on his PC. He thought for a moment when he was prompted for a file name. He typed *dory*.

DeWald rolled down his car windows to let the heat escape from his vehicle. Gauging by the internal temperature of his car, DeWald figured the day would be more summer-like than spring-like. He loosened his tie and rolled-up his shirtsleeves. He was leaving his third antiques shop of the day. So far none of the proprietors could recall selling any harpoons recently, but all agreed they would keep an eye out. DeWald left his business card. The thought of not having to enter another antiques shop for the rest of the day set well with him—one more Beanie Baby and...He picked-up the brown paper bag on the passenger side floorboard and reached inside, pulling out a plastic encased baseball card. On the card was Carl Yazstremski, Yaz for short. Work did have its advantages, especially when one could work his hobby and his case simultaneously. He had grown up in the Yaz era, even gone to a 1975 Series game. Those were the good old days—like the day when Pudge Fisk, standing in the first base line, used every ounce of body English he had to persuade his just hit long ball to remain in fair territory. Now it seemed money ruled and the game took second place. Still, the game had its mystique—a trip to Boston and Fenway Park sounded awful good right now. He would plan on it, he and Linda.

His pager went off. Recognizing the number as Laykin's, he flipped open his cell phone.

"Good news, DeWald," Laykin said, not waiting for the obligatory hello.

"What have we got?"

"The boot, we know whose it is."

DeWald's jaw dropped. "So quickly? We got lucky."

"We did just that."

"Let me guess, the guy wrote his name somewhere inside the boot?"

"Nope. Just as good though, we found a receipt crumpled up in the toe of the boot. It's got the guy's full name."

"I'm there," DeWald said, turning off his cell phone and starting the car.

He felt a cool breeze soothing his overheated mind. Like a thunderstorm bursting over the horizon in route to land baked dry by a severe drought. It seemed almost too good to be true. And what about that old rule? If it's too good to be true, it probably is. The concept began gnawing at his jubilance. Why leave a boot behind? How stupid could the suspect be? Criminals had done stupid things before, some of which he had witnessed. But still...he shook his head. No. No, until he found out otherwise this was a breakthrough.

Thomas greeted him at the door to the lab. "Sweet. We've got our man." He put his hand up for a high-five.

DeWald looked past Thomas to Laykin who sat on a stool with a phone up to his ear. While Laykin talked he pointed to a plastic bag with a crumpled piece of paper sealed inside. DeWald carefully pinched the bag at the top and lifted it to eye-level.

Laykin set down the phone and nodded at the bag DeWald held. "We're picking him up for questioning. Recognize the name?"

DeWald squinted. "Can't say I do."

"His name was on the NBWHS membership list. Before you call me a freak for remembering, recognize that I have a nephew with the same first name—"

"Let's get a search warrant—"

"Already in the works, it's with the judge even as we speak."

"It all seems to fit nicely. He's a member of NBWHS and therefore has motive. Has Detective Allbright gotten to know this man at all?"

"Supposedly."

"Does the Detective know we found this?"

"Just got off the phone—"

"And?"

"Doesn't think it's a fit."

"Why?"

"Says our suspect doesn't fit the profile."

"The way you said that leads me to believe you don't agree with the Detective."

Laykin shrugged. "I've been around cops almost all my life…"

"And?"

"And I know when one is becoming a buddy with a suspect. It happens all the time in these undercover operations."

"So you think the Detective is not being objective?"

"Does a bear crap in the woods?"

The Accord was spotted in the drive of the ranch style home colored in Nantucket gray. The patrol cars lined up on a street over a block away. At homes along the way window curtains were parted and mini-blinds gapped, as curious eyes peered out into the street. A S.W.A.T truck pulled in behind the patrol cars. Officers bearing arms congregated behind the truck, engaged with their radios. One man, wearing fatigues, held open a map he spread across the trunk of a patrol car and began pointing to areas on it. Groups of officers shot from the group and began running in a crouched position between homes.

Bennett heard the neighborhood erupt with the barking of dogs, and listened for the distant siren that normally sparked such commotion. But he heard nothing above the barking. He stopped listening and thought about working on dory, but at the same time found himself growing tired of it and contemplating why he should continue pursuing something so pointless. His nemesis had never done any real harm. Wasn't he above heckling and jeering?

By now those peering through the curtains had grown brave and moved out to the front steps of their homes. The officers ordered them back. One elderly man protested, but the officer pushed him back through his front door. The eyes returned to the curtains and blinds. Most of the officers had dispersed, the remaining ones crouched behind their squad cars, weapons drawn and aimed up the street.

The barking continued. Bennett got up from his seat in front of his PC and stretched. He looked out the window to the street below and was startled to see Alfred making his way up the sidewalk. At first he thought he would feign not being home, but realized his car was out front, and the thought of joining Alfred in his walk and getting exercise did draw some appeal. He started downstairs and met Alfred at the front door.

"You a mind reader?" Alfred said when Bennett opened the door before he had had a chance to knock.

"Saw you coming from upstairs. You taking your walk?"

"Stretching the old stork legs."

"Mind if I join?"

"Not at all. Can I get a glass of water first?"

Bennett led Alfred to the kitchen and handed him a glass of water. Alfred took a few gulps and tossed the rest out the kitchen window.

"Why'd you do that?"

"Odd habit of mine. Let's go."

Bennett outstretched his arm, tacitly directing Alfred to lead the way. The two stepped out in the front yard and headed for the street.

Bodies crouched behind bushes, moved in behind the two. Before Bennett and Alfred reached the sidewalk, both were pinned to the ground—officers lying on top of them.

"Damn!" Bennett thrashed. "What's going on!"

He overheard Alfred struggling and screaming for the brutes to get off him. Blades of grass, and the sod collected in their roots, crept into Bennett's mouth, as his head was shoved face down into the neighbor's lawn. In a split second, he thought it funny that his mind contemplated how much the neighbor's dog crapped in the yard. He felt a stab of a pain as his arms were drawn up behind his back. He panicked. Muggers? Yelling to his attackers to take his wallet from his back pocket failed to lighten the pressure on his shoulders. He could hear Alfred's moans and complaints, but not a word from their attackers. As he continued his pleas he heard an authoritative voice, yell, "Get off him..." They stayed put and an argument ensued between the man with the authoritative voice and some other man. It began to occur to Bennett that the present activity was not the work of hoodlums, but the actions of the authorities—exactly which ones, he had no clue.

What could I have possibly done? Bennett went through a mental exercise of reviewing the ledger of his latest activities, searching for anything he might have done that could be construed, however remotely, as criminal. But nothing came to mind. He heard one man yelling that they—which he knew to be himself and Alfred—were probably armed and that every precaution must be taken; to which the other replied that it was obvious they were unarmed. The argument ceased and, as if rocks were removed from his back, Bennett could feel the weight of his attackers being lifted from him. And then the argument continued.

"Ever heard of innocent until proven guilty. Damn, we're just bringing him in for questioning—"

"We can't take any chances."

DeWald held an angry stare at Laykin. Laykin looked down at the ground and shook his head. "Hold on to them tightly boys."

With officers holding each arm, Bennett rose to his feet. They were cops. Looking very confused, he looked to the man who had been yelling to help him and said, "I saw you at the beach the other day. You're with the FBI. What's going on?"

DeWald took a deep breath. "We would like to bring you down to the station for some questioning."

Bennett shifted his shoulders and winced. "About what? I haven't done anything. Is it a speeding ticket or something?"

"Nothing that simple I'm afraid. I think we should move this discussion down to the station."

"Do I need a lawyer? Can I refuse?"

"Whether you need a lawyer or not is up to you, if you've nothing to hide, why get one? As for refusing, yes you may, but I highly recommend against it. It wouldn't look good for you."

Bennett sighed and looked over at Alfred. Blood trickled from his nose and his balding head had green patches from scattered grass stains. "You okay, Alfred?"

"Yes, son." Alfred looked over to DeWald. "You need me too?"

"No, you're free to go. We may need to talk to you later."

Alfred looked Bennett's way. "It's okay son. I don't know what they want to talk to you about, but you're a good boy and I know you're innocent. Sorry to abandon you, but the wife should be home now. Don't hesitate to call if you need anything."

The two officers holding his arms pushed him in the direction of the squad cruisers and the three began walking. Bennett watched as all eyes focused in on him, from the neighbors whispering in their yards, to the uniformed officers standing behind their cars, to the plain clothed law enforcement personnel scurrying in front of him. The dogs ceased barking, as S.W.A.T. team members left their place of hiding for the street. Almost as far as Bennett could see, cruisers lined the road—red flashing lights chaining from one cruiser to the next, all within an overcast of eerie silence. A voice spoke behind him, asking him to confess, but the voice was told to shut up. Another voice hissed the word sicko. The patrol car door opened and Bennett was guided inside, an officer was pressing down on his head with stern control, and Bennett, his heart was thumping through his chest.

Chapter VIII

He sat across from the principal in a hard wooden chair that had numbed his hindquarters. His hands smelled heavily of onion, a smell he detested. The principal reclined in his chair behind his desk, his arms folded across his chest, and his eyes peering over the top of his half-lenses; he said nothing. Chucking the onions into the ceiling fan seemed like an interesting idea at the time. Setting half buried in a flatbed by the window of the second grade classroom, most were failing the experiment and weren't really growing anyway, so why not toss them into the fan and have them diced in mid-air? But the tears that were soon to well in Bennett's eyes weren't from the chemicals released from the chopped onions; they were forged from the principal's implement of discipline, a paddle that hung on the wall, in plain view, decorated in various colors of ink, signed by those who had experienced its crack. A young Bennett braced himself.

He felt much the same way when he found himself in the room with the detectives and the agent from the FBI. The only real difference was he knew he was innocent of whatever they were holding him for. They looked upon him in silence. The fat one, breathing with vigor, later introduced as Detective Laykin, the one he remembered from the news cast days ago, rapped the song, California Girls with his knuckles on the edge of the plain wooden table between he and Bennett. Chinese water torture? Did they think this would break him—trying to create a tension in the air so taut that if pressed against his skin would lacerate him, causing him to spill his guts. The thin, younger detective chewed his gum in rhythm with the knuckle rapping. He looked polished; his hair slicked back, suspenders, and gun harness over his shoulder. The agent leaned against the corner, his eyes moving between the two men at the other side of the table from Bennett.

Words were scribbled into the table. Most vulgar, informing the authorities to kiss parts of the writer's anatomy. A singular one, near the edge, just beneath Bennett's eyes, apologized to someone named Beth. The heavy man cleared his throat and asked Bennett if he owned a boot like the one detective Thomas had placed on the table. They were army green with a yellow tip over the toe area. Bennett wondered who Beth was—a girlfriend? Mother?

A hand slapped sharply against the table. Detective Laykin repeated his question, this time his voice fraught with anger. Bennett wanted out of the room badly, thinking that perhaps just outside the door of the room the nightmare might end. He had to hold at bay the urge to run for it. He looked at the dungy boot and answered that he did at one time own a pair very similar. Detective Laykin swished his lips and asked him why he qualified his response with *at one time.*

Bennett wondered who might be standing on the other side of the black plate of glass secured into the far wall behind the detectives' heads. Probably folks with profiling skills, one's who, with a keen eye, could pick up on the nuances of his movements and responses, and derive some conclusion as to whether he was guilty or not. Suddenly he felt very self-conscious, his movements becoming awkwardly staged. But how was he to act? What crime was he was trying to conceal?

He informed the detective that the boots he had that were similar were recently stolen from his car. The response from the floor was how convenient it was that this happened. Detective Thomas asked if this were so, why he had not reported it. The detective's eyes wandered to Detective Laykin, as if looking for reassurance that he was on the right track. Bennett shrugged, and said that the boots weren't worth much, that it hardly seemed worth it at the time. The Detective reached for his inside coat pocket and withdrew a baggie. He rested it on the edge of the table and flicked it with his finger to send it Bennett's way. Bennett leaned over to look at the baggie, squinting to look more closely at the crumpled white slip of paper inside. The reflection on the plastic from the bright florescent light hanging from the ceiling above made it difficult to see the printing. Detective Laykin offered to help and proceeded to describe the contents of the slip of paper, a receipt from a local tackle shop for 30-pound test fishing line signed by Bennett Ackerman. Bennett sat back in his chair, noting how white his fingers appeared while resting in his lap. He nonchalantly told his audience that the boots were probably his.

The rust colored stains in the white tile ceiling made Laykin's crap-eating-grin all the more pugnacious. He developed the smile just moments before attempting to shove words in Bennett's mouth concerning his contempt for Kathy Rueger. He described for Bennett the disdain he knew

Bennett held for Kathy Rueger, and why he would want her gone, gone in a permanent sense.

It dawned on Bennett.

He found himself laughing. Relieved to hear it was something he knew for certain he had nothing to do with. But the laughter was short lived when he realized the others were not joining him, that he wouldn't be here if they didn't believe he was in some way involved. He asked them if they were kidding. The agent in the corner moved forward and asked Bennett if he'd like a Coke. A Coke? Bennett told him what he'd really like was to just go home because this whole thing was a charade. Laykin advised him to take the Coke. He proceeded to provide Bennett an orientation—as if he were a new recruit at boot camp—on what Bennett could expect for the rest of the day, and, appearing to cherish the thought, added at the end of the sentence, that it might apply to the rest of his life as well. The process was nothing so trivial as a job interview, his freedom hung in the balance, and rather than having a great concern over whether or not an offer would follow, he must now contemplate whether his back might meet the cold steel of the table he could be strapped into just prior to receiving a pin-prick administered complements of the state.

Bennett agreed to the Coke after Detective Laykin informed him as to what he could probably expect in terms of duration for the day's interview. Agent DeWald stepped from the room. Detective Laykin leaned into the table, making his bristly, pasty face a matter of a few inches from Bennett's side of the table. He tipped his head in the direction of the door, in reference to the departed agent, and informed Bennett that if it weren't for the Feds hanging around he would be kicking Bennett's butt at the very moment. The force of the message was enough to make Bennett wonder if maybe he really hadn't done something.

Agent DeWald returned with the soda and placed it on the table in front of Bennett. Was it good cop, bad cop? He didn't think so; he believed the Detective meant every word he said. He looked over at Agent DeWald and proceeded to tell him that he was innocent, that it must be some big mix up, there was no way he had anything to do with the woman's murder. Detective Thomas stepped in and told him that if he cooperated and if he were telling the truth, than he had nothing to worry about. He was asked how he could explain his boot being present at a crime scene.

Easily. Whoever stole it from his car left it there.

Too convenient.

Katy Rueger—she spoke badly of the NBWHS; motive enough for anybody, both detectives agreed with a nod of their head. Detective Laykin added that he might be tempted to knock someone off if he had found himself in a similar situation. Bennett responded that he didn't hate Kathy

Rueger that badly, not enough to kill her. Thomas smiled and told Bennett that a jury will probably have to make that call.

Court is not a forgone conclusion!

The pounding of questions continued for some time. Bennett watching himself in the black mirror, just to see what *they* were seeing in him. How does an innocent man act? Detective Laykin got up from his seat and stretched and informed everyone that he needed to drain his lizard. He looked over to Agent DeWald and told him that Bennett was his for a while.

Agent DeWald sat down opposite Bennett and began some invisible doodling on the tabletop with his finger. He was not as convinced as the other two that Bennett was guilty of the crime. Agent DeWald explained that Laykin and Thomas were under the gun from their Chief and the pressure was perhaps getting to them and they may not be as patient as they should be in cases as serious as this one. Great, a shotgun trial. They were searching his house, even as the two spoke, agent DeWald informed him. The agent cracked his knuckles and asked Bennett if he would be willing to take a lie-detector test, to which Bennett wasted no time in answering that he would. Agent DeWald made it a point to let him know that if he lied during any part of the test, say hypothetically if he were trying to protect someone, that it could cast some doubt on the entire interview.

Protecting someone, whom could he possibly be protecting? And then for some reason he thought of Tanya.

As the wipers flew across the windshield, a lone streak of water annoyingly arced at eye-level. DeWald shifted his head down slightly to look below the obstruction. The guardrails from the Braga Bridge flew by the side window. Starry reds from taillights refracted on the wet windshield. At the other side of the bridge, DeWald pulled into a service station and parked near the back next to the restrooms. He sat for a moment and then went inside. The cashier remarked to him that they were in one hell of a storm. DeWald ignored him as he pushed down on the lid of a black carafe, pumping stale coffee into a styrofoam cup. As he poured a packet of sugar into the cup, the cashier spoke up again, suggesting it was the type of night where the New Bedford serial killer would strike. Now the city of Fall River was in on the hysteria.

DeWald stepped out in the cold rain and returned to his car. He did not notice the steam rising from his coffee and the way it completely fogged his windshield. He had interviewed Bennett Ackerman for hours and not really gotten anywhere. Every instinct said he was not the man. What little profiling training he received at Quantico gave him no reason to point to a man with a lifestyle such as Bennett Ackerman. But the physical evidence was piling up: the whaling paraphernalia in the home, the boot, the files on the PC. DeWald shook his head; the files were extremely hard to explain, particularly the image file of Sheryl Bidwell's dead body and the files of the strippers Patton used on his web site, including a picture of the Bidwell girl. Maybe the guy's a good actor.

A car rolled up beside DeWald. He listened as the vehicle's engine stopped and a car door shut and watched as his passenger door opened. The occupant wore a wind parka with the hood up. The hood was pulled back and long auburn hair flowed beneath it.

"Detective Allbright, a.k.a. Tanya Appleton?" DeWald asked.

Tanya ran her fingers through her hair. "It's Stacey, Boston PD."

"How did you get stuck in this mess?"

"They needed an outsider—"

"An attractive outsider—"

"Please," Stacey said, rolling her eyes.

"Sorry," DeWald said sheepishly. "Glad to finally meet you."

"Likewise, Agent DeWald." Stacey avoided the fogged windshield and looked out the passenger window back at the bridge. "Actually I volunteered because I see this as a challenge and because I have some personal business to look into while I'm here."

"Oh?" DeWald didn't get a response. "I won't pry." He sipped his coffee. "Is it safe to say you're upset by all this—"

"He's not your man." She shook her head, as if denying it all. "You guys jumped the gun."

DeWald could see intensity in her face. "I think we agree on that point."

"Damn rain..." She shifted in her seat. "I've spent some time with him and believe I know him fairly well. He couldn't have done any of the killings."

DeWald tried to be sensitive. "Yeah, I'm sure you're right, but is it possible he could have some mental disorder? It's happened before—I've been on cases—"

"He's as sound as they come, DeWald. I'm only aware of one major injury in his life, emotional or otherwise, and the only side effect was loss of memory. And I'm no specialist, but I don't think head trauma can cause violent mental illness."

"Does he still experience memory loss?"

"I don't think so. He remembers everything we've done together."

"What about this theory: he commits the crimes, blacks-out, and when he awakes, no longer recalls what he's done."

"That's ludicrous." Stacey pulled a kleenex from her purse and cleared a spot in the passenger window. She looked out, away from DeWald.

"It's happened before." DeWald waited for a reaction, but received none. "What about all the whaling pictures on the walls?"

"Simple. He acquired that passion from his father. He loved his dad dearly; I think it's his way of perpetuating his father in his life. Harmless, really."

"So his father is dead?"

"Died when he was a teenager."

"Natural causes?"

"What does it matter?"

DeWald took a sip of his coffee. The attendant peered around the corner at the two and grinned. "I think he thinks we're necking..." Stacey didn't laugh. "There could be some guilt associated with his death, that often brings about aberrant behavior."

"Not violent behavior."

"Sometimes."

"Look neither of us is trained in this area, we shouldn't even be speculating—"

"How did he die?"

Stacey sighed. "Accidental drowning..."

"And?"

"He wouldn't talk about it. Had to research it myself."

DeWald nodded. "So there are definitely feelings of guilt."

"Who knows? But even so, it doesn't manifest itself in violence."

"You're very sure about how well you know this man aren't you?"

Stacey leaned her head on her hand and rested her elbow on the door handle. "I don't know. I mean, I guess it's my intuition. But I've found my intuition is very reliable. I inherited that skill from my father."

"So you're father was a cop?"

Stacey smiled. "The best." She sat up. "He's retired now."

DeWald caressed his chin. The rain picked up and noticeably began pounding the car top. "Look," DeWald said, "what about the other members, anybody else look like a potential suspect?"

Stacey removed her head from her hand and began tapping her finger on the door armrest. "I don't know the others that well, but they appear on the surface to be harmless."

"Have you spent time with anyone other than Bennett?"

"Some—"

"Some! Damn—" DeWald caught himself and took a deep breath. "When are you going to get to know some of the others?"

"Soon. You know doing this kind of work takes time. You guys are jumping in here way too early. You're not giving me the time it takes to do this job correctly. It's almost an art you know. You can't just bulldog your way in an expect to get results."

"Time is running out."

"So you and I will stand by as they hang an innocent man?"

DeWald turned to face Stacey. "Shoot straight with me, are you really going to get to know the others?"

"*Yes.*"

"So you'll start spending less time with Bennett?"

"Yes. I'm going with him to visit his grandpa's tomorrow. After that I'll start pulling away."

"Don't fool yourself into believing you can have a relationship with this guy. You know as well as I do when he finds out you're a cop working undercover, he'll drop you."

Stacey shook her head. "I never allow myself to seriously consider a relationship. This is work. But I do find my self caring for him."

DeWald nodded. "I'll buy you whatever time I can."

Stacey put back on her hood. "Thanks, got to go."

DeWald grabbed her arm just before she stepped outside the car into the rain. Stacey looked back coldly. "It's for his own good...you know that...you catch the real guy and Bennett's free and the only way to do that is to spend time with the real killer."

"I'll do my job," Stacey said, as the door shut.

DeWald looked at the Yazstremski card resting on the console. "I should have played baseball, life would be so much simpler."

Bennett's Accord was beyond Interstate 195 and on highway 6 at the rotary to the Bourne Bridge, before much was said between he and Tanya. The early morning brought a light fog that clung to the ground. Despite his affection for her, the files he found and the questioning by the police consumed what thought processes he had available. He knew he was potentially in very serious trouble and couldn't help but mull over what he could do to exonerate himself. He knew an arrest was pending and that he needed to come up with something before it happened. There was no way he could make bail, not for a capital offense.

"So tell me about your grandfather?" Tanya slipped in.

Bennett heard Tanya say something, but didn't catch any of it. "Huh?"

"Your grandfather, you've told me so little about him and here we are about to meet him."

"Oh…yes, Granddad," Bennett said struggling for something to say, "well, he's a great guy. When I was a child I spent a lot of time at his place on the Cape. He always took good care of me. You know, like only a grandparent can. My Dad always said I took after him."

"Sounds like a great guy, and yes, you must take after him."

Tanya turned her head towards the passenger window. Would she shake her head on the witness stand, Bennett thought, denying she knew much about him, covering for herself, or whomever else she might be working with? And just how much was she really involved? How could she possibly have gotten those files without knowing something about what was going on? She turned to look out the windshield, her eyes intent on the road. So the girl in the image was murdered, and by the same hand of the one who murdered Kathy Rueger. Perhaps she was jealous of these women. As Bennett pondered the thought and similar ones, a curtain of alienation unfurled in the inches between he and Tanya. Who was she? Could she hurt him?

Tanya turned from the windshield and looked at Bennett. She was having a hard time acting as "Tanya." Stacey kept slipping through. He looked pale and cold and at a great distance. Physically driving the car with his mind continents away. What was he thinking? Was he capable of killing? How did she really feel about him? He moved little, save for his hands and his eyes. His eyes peering down the road, oscillating with thought, but what? she thought. She knew he didn't know she knew, but should she tell him? She thought back to day one, when she knew she was destined to become a detective; father was one. She never wanted anything else. But to tell Bennett, to help him, would entail a violation of the trust afforded her as a detective—something she felt very strongly about upholding. Besides, she would be removing herself from the social neighborhood of the killer. She instinctively felt the killer was there, living and breathing in the NBHWS community. To remove herself would be to greatly lessen the ability of the entire force to catch the killer. Others could be murdered and she would be at fault, not directly of course, but indirectly, because by staying in there, she might have intercepted him on his rampage to the next victim. Bennett kept his motionless pose. She turned her head towards the passenger window. He was on his own; it had to be that way.

Bennett cracked the window and allowed the salty air of the Cape to whistle through. Somebody was setting him up and he was inadvertently assisting. The search warrant had been executed; he knew that. The laptop was gone. They would find the files. How could he explain he had them? Blame it on Tanya? He thought on it for a moment and then quit; he couldn't do that. So instead he would die by lethal injection? Maybe he should confront her, ask her how she got the files. She looked so innocent and charming in the seat beside him. Asking her may blow everything; destroy the scaffolding of the relationship he had so diligently built with her. Better maybe to wait. Things would get cleared up; they had to. He was innocent. Why was he worrying? He was innocent. I'm innocent!

"Kind of a dreary morning..." Bennett said, forcing himself to mentally drop down a gear.

Tanya rubbed her forehead. "Yeah. I think the sun is supposed to burn through this and we should have a good day...weather-wise."

Bennett smiled. "Weather-wise? I think all-wise." He reached over and grabbed her hand and gave it a gentle squeeze.

Tanya grinned. "You're a very sweet guy Bennett. Always know that I think of you that way no matter what happens."

Bennett shrugged. "Okay...although only good things are gonna happen."

Tanya cocked her head. "Life isn't always fair."

Bennett widened his eyes in an exaggerated fashion. "Whoa...heavy."

"I'm serious, Bennett."

Bennett breathed a sigh. "Do you know something I don't know?"

"Just had some experiences in my travels, that's all."

"Old boyfriends? Work? Family?"

"All of the above. Sometimes fate forces you to make a choice between different parts of your life. You can't always have your cake and eat it too."

"I suppose that's true," Bennett said, nodding without giving the words much thought. He heard what she was saying, not the lines themselves, but what was between them. Historically, words like hers, as spoken by previous female acquaintances, had been the kiss of death to a relationship. He couldn't contemplate it now; one more serious thought on his stack of thoughts would cause his entire mind to blow. "Granddad used to play tennis with me," he said, reverting to the earlier conversation.

Tanya paused in her response, deliberating whether to allow the conversation to take a turn. She sniffled. "Did you beat him?"

"Heck no. I was terrible. He beat the tar out of me. But you know what? He never lost his patience with me. I was a spaz, but he never gave up on me. Of course in the end, I gave up on me. I couldn't play worth a darn. I was really, really, bad," Bennett chuckled.

Tanya nodded. The fragrant odor of Honeysuckle skidded through the cracked window. "That smells wonderful," she said.

"Honeysuckle. The entire backyard at Granddad's house is filled with the smell. It grows all along the fence. Almost boggles the eye."

Tanya took a deep breath. "It's nice not to be at work today. Work can be so draining. Of course it depends who you work with. What about at your office, is it stressful?"

Bennett shook his head and pursed his lips. "We get along real well. I really don't consider it work."

"What about some of the guys? What are they like—the ones closest to you?"

"Geeks, like me—"

Tanya pushed Bennett on the shoulder. "You're not a geek," she said. "Seriously though, any oddballs?"

"Haven't you asked me this before?"

"Can't I ask more than once?"

"Well there's always Elander..."

"What's so strange about him?"

Bennett cocked his head. "What's strange about him? The question is, what's not? He's into this retro stuff—that stuff out of the seventies. You know, smiley faces, bell-bottoms, shag carpet, Crosby, Stills and Nash, Quisp cereal—all that stuff I love to hate. I think of him and I think of a powder blue leisure suit."

"So he's a peaceful type, doesn't like violence?"

"To the extreme—"

"It's often those guys—we had a guy like that in our office—who flip out. I'm sure Elander is capable."

Bennett pictured the curly headed Elander popping into the office one morning wearing army fatigues under a turquoise sweater vest with a plastic belt buckle, and a M-16, spraying the halls with bullets and yelling obscenities as hot lead ripped through the bodies of his stunned coworkers. But this picture was quickly distorted by a song that played in his head, perhaps some remote connection to war..."Teach, your children well..." He shook his head. "I don't think so. He literally would place a foot in a puddle to sidestep an ant."

Bennett observed the antique shops lining the road. Window after window displayed pallid translucent Depression glass pieces sparkling in the sun. "We're in Sandwich. Lots of glass was made here."

Tanya squinted as she peered into a passing shop window. "I've never cared much for glass stuff."

"My mother collects it," Bennett said. He looked in the rearview, focused ahead, and returned to the rearview. "As a child, every once in awhile Dad and I would hop in the car and drive up to Sandwich to pick up some glass pieces for Mother." As he thought of it, he could smell the musty scent of the shops. "Neither of us had much of a clue when we started, but just before his death, Dad was becoming quite the expert and I wasn't far behind. Mom was a tough customer—we always knew when we picked a bad piece; it mysteriously ended up on a dusty shelf in the basement. It's funny though, in an odd way, after Dad died, the pieces were polished and moved upstairs into the display cabinet with the good pieces."

"Death makes us rethink the value of most everything."

"Yeah, I guess that's true. After Dad died, nothing seemed of much value. I'd have given everything I own, to have him back again—you know, just like the song."

"Yeah…" Tanya pulled a pair of coral framed Ray-Bans from her purse. She flipped her head back and slipped the frames on.

Bennett felt his left leg falling asleep and shifted in the seat. "We're not far away from Granddad's."

"So how do you feel about your Dad's death—I mean today?"

"I miss him."

"Do you feel guilty?"

Bennett nodded. "Always will—I'll never escape that. It's not something you easily forget."

"You've never said much about your mother…"

"What's there to say? She's very normal in every way. She was a housewife turned recluse when Dad died."

Bennett turned left on highway 149 to West Barnstable. Salty hay marshes sprawled on either side of the road. The sun burned bright, the fog now long since dissipated. A strong wind blew in from the north shore of the cape. A man struggled to keep the canvas roof on his fresh vegetable stand from kiting.

"Hope Granddad got some of that fresh Silver Queen corn," Bennett said, watching the man drive a stake into the ground.

Adorned in a quaint patina of unpainted weathered shingles, the cottage rested on a hill overlooking the north shore of the cape. Opulent pink country roses climbed a thatched archway marking the way to the front door. A mock orange tree shaded a side of the garage, slightly jutting from the cottage proper. Peonies painted in purple chiffon lined the cottage from the mock orange to the roses. Gold Flame honeysuckle climbed thatch set snug against the shingles.

"See him," Bennett said, pointing at something on the honeysuckle.

"See what?" Tanya said, her eyes hidden behind the Ray-Bans.

"He's right there."

"Ahh…" Tanya said, smiling. "A hummingbird, right?"

Bennett took a half step closer to the bird. "I was hoping we'd get to see some; they're such an unusual bird—"

"Welcome, welcome, welcome!" shouted a tall white-capped man with vine legs stepping out from the front door under the rose-blanketed archway.

Bennett held a large grin, looking up at the stale wiry man donning a gray sweater vest over a yellow plaid shirt and blue jean shorts. Stout black-rimmed glasses barred over his roseate covered cheeks.

"Tanya, meet Granddad," Bennett said, directing his hand at the man.

Granddad gave Tanya a hug and moved on to Bennett. "So glad to have you here, son!" he said, and then looking back at Tanya, "so glad to have such a beautiful young lady here too!"

The discussion then fell to the mundane drive up, the garden, and the weather, before everyone stepped inside.

"I've got some home brewed ice tea, if that sounds good to you," Granddad said, once everyone was inside the vestibule.

Tanya removed her sunglasses. "Sounds good."

"Why don't you two join Patrick out back and I'll bring you out two ice cold glasses of it."

Bennett hesitated.

"It'll be alright," Tanya said reassuringly.

"Here goes nothing," Bennett said under his breath.

The two made their way down a half-floor staircase to a family room. Tanya grabbed Bennett's arm as he reached for the handle on the sliding glass door leading to the backyard.

"Is that you?" she said, pointing at a photo atop a bookcase, of a man squatting with two boys standing at either side of him, all holding fishing poles. In front lay a row of silver fish that sparkled and were sunken in the grass.

"The one on the left with the buzz?"

"Yes."

Bennett grinned. "Yeah, that's me—dad is in the center and Patrick's on the right. We hit a school of blues that day. As you can see in the photo we caught a ton. What a day of fishing that was."

"You're cute. Your dad was a very handsome man."

"He was the best." Bennett pulled on the handle to the sliding glass door. "Shall we?"

Patrick attended a waist high grill placed near the back of the patio. A glass round table with an umbrella top stood in the middle of the patio. Four chairs with liberal blue and white stripes were placed around the table. On the table a basket of salt and vinegar chips had slid dangerously close to the edge. Tanya nudged them back. The smoke from the grill flew horizontally towards the cottage, blown by the strong winds off the sea. The umbrella flopped in the wind, vibrating the entire table. Patrick held a spatula by his side; his head pointed to the beach, his body not moving.

"Patrick," Bennett said.

Patrick maintained his stiff posture.

Bennett cupped his hands around his mouth. "Patrick!"

Patrick slowly placed the spatula on a small wooden platform on the side of the grill and turned around. He pulled his wind tossed hair to the side with his hand. A forced smile came to his face. "Bennett."

Bennett awkwardly walked up to Patrick and shook his hand.

"Is that the best you boys can do?" Granddad's voice rang out behind them. He placed the glasses of tea on the table. "Come on...give each other a hug."

The two men, turned temporary actors, managed a modicum of a hug. Bennett turned and motioned for Tanya to approach. He introduced Tanya to Patrick. A smirk crossed Patrick's face. "What's a beautiful girl like you doing with a dork like Bennett?"

Granddad spoke up. "Patrick."

"I'm just joking, Granddad." Patrick turned to face the grill and then turned back around. "So how's the ISP business, Bennett?"

"Not bad," Bennett said guardedly, "we're recovering from the winter."

"You still doing that programming crap—"

"Watch your language, Patrick," Granddad said, now resting in one of the striped chairs, his legs crossed and his top leg bobbing. "This wind is gonna leave me bald."

"Not much anymore. Too many other things to take care of," Bennett said.

Patrick turned to face Tanya. "This guy used to spend all day in front of his computer when we were out here at Granddad's. He'd cry until his dad would agree to take the computer out with us. Look around at the beautiful beach. Can you imagine, sitting inside with a computer, instead of spending time at this glorious beach?"

"—That's not exactly true, Patrick. I spent some time outside."

"Very little."

"Where's Candice?"

"Out spending my money somewhere I suppose. She didn't have the time for this, so screw her, you know. I'm not going to beg her to come, you know." Patrick grinned and tilted his head in Tanya's direction. "So Tanya, what do you do?"

Tanya explained her line of work and Patrick pretended to listen, his eyes combing the terrain of her body. Bennett walked back to the table and sat down by his granddad. He picked up his ice tea and took a sip. "He's never gonna let me forget what happened."

"It's his problem son. I imagine someday he'll grow up." Granddad took a deep whiff of air. "I'll tell you, there's no sweeter scent then the smell of the ocean..."

Good idea, Bennett thought, let's move on to another subject. "I know. I wish it could be bottled."

"So what do you do, Patrick?" Tanya asked, folding her arms across her chest.

Patrick smiled and picked up a beer resting next to the spatula and held it up in front of Tanya. "Own a bar."

"I hear you own a boat too."

"Yep. A sweet little toy."

"Bennett told me you had a nice boat. Probably fish a lot with it or something..."

Patrick got a cocky smile and laughed. "So happens I do fish with it. I've caught some freakin' monsters—twenty, twenty-five pound blues—"

"Some whales, huh?"

"Huge..."

Bennett was looking over at Patrick. "So Granddad, still playing cribbage with the boys down at the lodge?" He couldn't hear what Patrick was saying to Tanya, but he didn't like the sardonic look on his face. In his mind, the look and Patrick went hand in hand, one in the same, synonyms— the same look Patrick had given him one day when they were much younger and attending church where communion was served and, while holding the body of Christ in his hand, a piece of saltine cracker, he had asked, *Where's the peanut butter?*

"Nah, getting to be a little too far to drive these days. Spending more time at home, with the garden and all. You know as you get older, simple things in life, such as gardens, become more of a draw for one's time."

"I can tell—it's beautiful. You're really good at it."

"Got my gardening skills from your grandmother."

"She must have been very talented."

"Talented..." His still eyes disclosed his drifting thoughts. "Yes, she was good at just about everything she did."

Bennett thought of her and realized he didn't know much about her, even the most basic information. Why he started with the basic information question he did, he wasn't sure. "How'd she die?"

Granddad took a sip of his tea and sat back in his chair. "She had a strange thing come over her a year or so before her death. I'd say I was about 30, so she must have been about 28. Anyway, she started to

155

experience blackouts in her life. She'd do things, sometimes half crazy, and not recall that she had done them. Doctors thought it might have had something to do with a head injury she had as a child. Fell out of the buggy when she was four. Of course doctors back then weren't as sophisticated as today's doctors are, so they might have been wrong. I always thought it might be genetic, as I recall her Daddy had similar problems. Anyway in answer to your question, she died in her sleep one night…"

Patrick gave Tanya the once over. "So why don't you drop Bennett and take me for a spin?"

She grinned. "I don't go for fisherman."

"Are you saying I'm fishing?"

"No, but if you are, your worm is too small."

"Ouch!" Patrick yelled out.

Tanya turned and walked back to the table. She sat down by Bennett. He looked at her with a puzzled expression. She explained his cousin was an overly sensitive man. Bennett nodded, acknowledging the sarcasm.

"Hey Benny remember this?" Patrick said, while pointing his finger towards the ground.

Bennett gave him a puzzled look.

"Don't remember, huh? What about this," Patrick pointed at his leg.

Bennett returned a blank stare that soon was overtaken by a wide grin. "Oh yeah, shoot to injure!"

"And this," Patrick said as he pointed to his head.

"Shot to kill," Bennett replied.

"Now what was the first one?"

"Take cover."

Tanya got a puzzled look on her face. "What in the world are you guys talking about?"

"Oh, it's stupid, Tanya," Bennett said.

"*So*—what is it?"

"Patrick and I developed these hand signals when we used to play war on the beach with the neighbor kids."

"We had a system—had them completely fooled, you know. Jimmy Sanches thought we could read each other's freakin minds," Patrick added, laughing.

For the first time Bennett saw some life come into Patrick. Maybe there wasn't such a gap between them.

Tanya wrinkled her forehead and smiled. "*Benny?*"

Bennett shook his head. "Don't go there. I don't like being called that."

"Okay…*Benny*," she said.

Bennett rolled his eyes.

Patrick brought the burgers to the table. Granddad had already put out a full plate of steaming Sliver Queen corn on the cob and a bowl of potato salad. Bennett held warm memories of his Granddad's potato salad—a recipe that actually came from Grandma's files, but Granddad maintained the tradition.

"You still hanging out with Paul?" Bennett asked, looking at Patrick and shoveling a large portion of potato salad in his mouth.

Patrick pried a buttery corncob from his teeth. "Was until last week."

"I can't believe the guy's still walking. That guy pumped more drugs in him than food. Remember the time he was really stoned and he drove his boat into the causeway?" Bennett chuckled and noted the rise it brought out in Patrick. Bennett looked over at Tanya. "You know what his first words to the cops were?"

Tanya shook her head, watching Patrick out of one eye.

Bennett slurred his speech saying, "Who put this damn thing here! It wasn't here yesterday," and broke out laughing.

A cold glare fell over Patrick's face.

"He was such a moron—"

"Knock it off, Bennett, the guy is dead, give him a break, okay!" Patrick yelled, the veins on his neck bulging. He dropped the corncob to his plate.

Bennett's jaw dropped. Granddad turned to face Patrick. Tanya eyed all three.

"Jeez, I'm sorry Patrick. I didn't know."

Patrick spoke calmly. "You think you're so much better than everyone else."

Bennett took the remark in silence.

Granddad ended up asking the obvious question. "How'd he die, son?"

"Boating accident. He was wasted. Fell overboard and I didn't realize it. Went like that," Patrick snapped his fingers, "they still haven't found his body yet. Stuff happens."

"How can they be sure he's dead?" Tanya inserted.

"You tell me. We're fishing late at night. One moment I look back and he's sitting on the stern with his fishing pole and ten minutes later when I look again he's gone. I turned back but couldn't find him anywhere. The toothpick sank."

"He didn't yell out?" Tanya asked.

"What do you think? If he had, do you think I would have waited another ten minutes before turning around to pick him up?"

157

"He'll surface somewhere," Bennett said.

"Don't count on it, the sharks more than likely got him."

Tanya scrunched her face.

"You're taking it pretty well," Bennett said, observing Patrick's emotionless account of the incident.

"The guy's dead, what am I supposed to do? Cry? Will that bring him back to life? What can I say, life sucks and then you die. Right?" Patrick took a long swig of beer.

"Life's what you make it, Patrick," Bennett said. "If you make it into crap, well…"

"Take a hike, Bennett."

Granddad grabbed his own left arm and grimaced.

"You never would listen, Patrick," Bennett said.

A gasp perforated Granddad's tightly clamped teeth.

Patrick leaned into the table. "I'll listen when you accept the fact that—"

"Guys!" Tanya screamed out.

Both looked over at her startled.

"Look, there's something wrong with your Granddad!"

"I'm okay," Granddad said, his face ashen.

Patrick's eyes dilated. "Granddad! Let's get him to a doctor," he said his voice shaking.

"It's okay, Patrick. If someone could just get my medicine—"

"Where is it?" Patrick asked, already at the sliding glass door.

"Bathroom cabinet, top shelf, only prescription drug on the shelf."

Patrick disappeared through the door. "He's more vulnerable than you are aware, Bennett," Granddad said, still holding his arm.

"To heck with him. We need to get you to the hospital."

"The drugs will take care of this, you don't need to worry."

A mosquito landed on Granddad's forehead and began draining blood. Tanya brushed it aside. It left a trickle of blood.

"He's bleeding!" Patrick yelled, his hands trembling so badly he couldn't manage to open the drug bottle. Bennett grabbed the bottle from him and opened it. Tanya held Granddad's head forward as a pill was popped in his mouth and his glass of tea tilted to his lips.

"Give it a few minutes," Granddad said.

The tear streaking down Patrick's cheek did not go unnoticed by Bennett. He patted Patrick on the shoulder. Patrick pulled away.

Bennett looked up from Granddad at the other two. "Let's get him to the hospital."

Chapter IX

The young man wearing the T-shirt and threadbare jeans moved back and forth between the three machines in the office. DeWald sat outside with Laykin and Thomas. They watched the man shaking his head and talking to himself. Getting the computer from the Mendosa home had proven easy; getting Patton's machine had proven difficult. Patton fought it as was to be expected—he was running live porn through it. A subpoena later, both computers were in the office. The third machine, Bennett's, had been grabbed during the search of his house.

Laykin was shaking his head. "So this is a computer forensic specialist?"

DeWald shrugged and said nothing.

Thomas rubbed his forehead. "Looks like some bum off the street. Hell, we didn't have to go clear to Boston to get one like him, there's a homeless guy down on William that would have fit the bill."

"The kid is supposed to be good. Let's wait to hear what he has to say," DeWald said.

"When we gonna arrest Ackerman?" Thomas asked, pushing a stick of gum in his mouth.

Laykin shifted in his chair. "I say we do it now."

"It's premature. If he's not the one, he can help us out considerably by working for us. Don't forget, he passed the lie-detector test," DeWald said, staring at the office where the young man was operating on the machines.

"Have you seen his medical records? He's got TLE for crissake! I've dealt with those types before, their mind is screwed-up—it allows them to believe their own lies. TLE people don't even know what they're doing half the time," Laykin said, shaking his head. "We found a printout of one of the murdered women, in his home for crissakes—and in it the tattoo on her forehead. She was obviously stone cold dead! How do you explain that? If he's not the one, then no one on the face of this earth is. What do we need? A written confession?"

"I know better than to jump to a conclusion and screw this case up so badly that we never get to the bottom of it," DeWald said, despite the damning evidence.

"You call it jumping to a conclusion, I call it saving young women's lives." Laykin picked himself up out of the chair. "You're playing with fire."

"We've got him under surveillance. What can he do?"

"Oh? Where is he now?"

"West Barnstable, visiting his grandfather."

Laykin tossed a breath mint in his mouth and rolled it around his tongue. "This computer forensic stuff is a waste of time too. That strange fellow in there isn't gonna be able to tell us anything Thomas didn't already discover through his interviews."

"Maybe not." DeWald reached his hand out and Laykin dropped the breath mint roll in his hand. DeWald dislodged one and sunk it in his mouth.

The young man stepped from the office and began walking down the aisle.

"Where you going?" Laykin barked out.

The man stopped and turned around with a startled look. "Getting a Coke?"

"What have you found?"

The man shrugged. "Well, I've searched all the files and found nothing the three shared in common."

"So in other words you've found nothing?"

"No, not true." The man adjusted the glasses that had slid down on his nose. "I found something in common in the cache."

"Cash? What does that have to do with computers?"

The man looked puzzled. Thomas spoke up. "Allow me. The cache on a computer is a temporary repository that holds previously used data for faster retrieval should it be required in the future."

"Thanks, Einstein."

"So what was in the cache?" DeWald asked.

"Well, both the girls were involved in web-based chat sessions."

"And…"

"And, they both chatted with someone by the screen name of Mocha Dick."

"Mocha Dick? What kind of name is that?" Laykin asked, the breath mint involuntarily tumbling from his mouth to the floor. "Damn it!"

DeWald watched as Laykin bent over and scooped up the mint and returned it to his mouth. DeWald winced and said, "Are you sure you read the name correctly? Could it have been Moby Dick?"

"No, it definitely read Mocha Dick."

"Moby Dick would make some sense."

Laykin rolled his eyes. "Like I said earlier, no help."

"Bennett's name is in the book, you know, Moby Dick that is, I checked," Thomas piped in. "It could mean something."

"Junior, go get a smoke, nicotine withdrawal is creeping in."

"I did find one odd thing on the Patton machine," the kid said as he folded his arms. "I found a file named *harpoon.* And if that weren't unusual enough, the file does not belong with the operating system on the machine, it's a Unix program file. It's a real mystery as to why the machine would have that file."

DeWald pursed his lips. "Best guess?"

The kid slid his hands in his jeans pockets. "Well, it could be that somebody downloaded the file from the Internet, thinking that they could use the file for something. When they found it wouldn't run, they just left it there."

The candy rolled about in DeWald's mouth. "Plausible, but I don't think so. These killings have an overall whaling theme to them. The file name suggests it fits in with the theme—"

"But the file is worthless to the machine. It's like you buying a book from the store written in Japanese, what good is it to you?"

DeWald looked at the kid and nodded. "Maybe whomever retrieved it, in this case the Patton girl, was naïve to that, or hadn't thought it through completely. Or maybe someone from the outside placed it there. Can you tell what it does?"

The kid shook his head slowly. "I'm not a Unix head, and even if I were, the file is binary—it would take some doing to untangle its instructions."

Thomas looked concern. "Is it a virus?"

The kid shrugged. "Hard to say, but as a general rule, Unix is not susceptible to viruses like other operating systems."

DeWald squinted. "Why do you ask, Junior?"

"Detective Allbright asked to review some of the files we found on the Patton machine. I think that *harpoon* file might have been on the floppy I provided her."

"It won't run on her machine," the kid said, shaking his head, frustration creeping into his voice.

DeWald lifted his hand. "Whoa…whoa…didn't the file name strike you as odd at the time?"

Thomas shrugged. "At the time, I guess not—"

"This is all of no help," Laykin barked.

DeWald shook his head. "I disagree. We now have another angle to this case."

"What's that?"

"There's an Internet element. Someone is remotely moving files around, and using the Internet to do it. And the chat sessions have some potential as well. Did you find anything on the other machine?"

"I didn't find anything more than what was previously reported, but then many of the files are encrypted, so who knows?"

DeWald shrugged. "Well, let's decrypt them."

The kid stifled a laugh. "Surely you're joking. They probably have a bit key longer than a whale's yang."

DeWald remained stoic.

"Can I get a Coke now?"

"Yes, but think about how we can get into those encrypted files."

He nodded, and took off down the hallway, and having turned a corner, burst out in laughter.

A damp cool breeze pushed the blinds out from the window, and when finished, allowed them to fall delicately back into the window frame. Softened light, from an all but full moon, when filtered through the blinds, striped the wall with floating alternating rows of gray and mint green. A steady rain filled the distance with white noise, while tapping on the gutter outside the window. It was a perfect night for sleeping. Bennett stacked his two pillows and positioned them against the headboard. He pushed himself up so that his lower back met the bottom of the pillows. He worked at focusing his eyes on the striped wall while taking a deep breath and pulling his hair back over his head with his hand.

The clock read 3:12 am. He had come to the realization, after hours of tossing and turning, sleep was not in his schedule for the night. He found his mind hopelessly rambling through the worries of his life. He thought about Granddad, Patrick, Tanya—the list went on. After the trip to the hospital, Granddad was back to being himself. The doctor admonished him for failing to take his medicine. He, of course, blamed it on the excitement he felt for the upcoming visit and swore it would not happen again. A tear, capturing the light of the moon, streaked down Bennett's cheek. The thought of losing Granddad...He got up from bed and walked into the bathroom where he wiped his cheek on toilet paper.

Who was Tanya? He knew enough to know he didn't know. Patrick was as distant as ever. He obviously still blamed Bennett for his uncle's death. Hell, he blamed Bennett for the entire sorry state of his life. Things were crumbling about him. And throw the imminent arrest into the mix…too much, too much to think about all at once. You're in it deep now, Bennett. Real deep. How are you going to get yourself out of this mess? His rational thinking was drowning in a sea of confusion. And when he panicked, and somehow managed to shoot his head out of the water and gasp a quick breath, he found some other problem drawing him back into the abyss. He took three deep and prolonged breaths, trying to relax his body. He knew he would need to address them one at a time. But which one first? It had to be his arrest. Granddad and Tanya would have to wait, unless Tanya was somehow involved, in which case he would keep her in the mix. He mentally threw Patrick out his bedroom window, the lowest of his priorities. Next, he would need someone to talk to…maybe Collin? No, Collin was a good friend, but would he be there through thick and thin? After some thought, Bennett concluded probably not. This needed to be somebody whom he knew would stick by his side regardless of how troublesome the situation appeared. Who could that be? The round face of a girl with wet brunette hair, licking a Popsicle after a swim at the beach, while leaning against the cement wall of the causeway, popped into his mind. He smiled. Hoosier, with her inviting smile, would be the one.

After a moment trying to determine the thumping sound, so awkward to the room, Bennett finally tuned into his own heart. The covers were tossed aside and his feet planted on the carpet at the side of the bed. He clicked his side table lamp on and vigorously rubbed his eyes from their dormant state. He dragged his hands along his bare legs while attempting to determine what should come next. Looking from his bedroom to the staircase leading to the second floor, he shrugged, pulled his robe from the back of the bedroom door, and headed upstairs to his PC. As Collin would say, time to go hog wild.

He went to the newsgroup first. Mocha Dick had replied.

There she bloooows! (LOL)

By the way, I should explain LOL since you're a neophyte—it's short for Laugh Out Loud. And so to, Mocha Dick is not Moby Dick, as you so obviously concluded—perhaps you require stronger reading glasses. I presume from the language of your article that you consider yourself an Ahab type. Just the same, you are not Ahab, for Ahab is my master. And should you be, though you aren't, I quote the next paragraph from your literature:

"…Ahab stooped to clear it; he did clear it; but the flying turn caught him round the neck, and voicelessly as Turkish mutes bowstring their victim, he was shot out of the boat, ere the crew knew he was gone."

163

The damned are in the belly! Mocha Dick carries the damned for the ambergris. Look to me, the rogue whale roaming the depths of cyberspace, to find the damned!

Mocha Dick (and, I'm Ahab too, surprised?)

Bennett pulled his lower lip back over his teeth and shook his head. *My clown is back.* Ahab or Mocha Dick, whichever the writer claimed as a title, which in and of itself indicated some sort of mental disorder, replied to a newsgroup based on the mysterious whale file. Earlier he wrote of giving the keys to the kingdom. Could the *whale* file have been the key? If so, then he knows something of the *whale* file, and perhaps even created it. He could very likely be the killer. And he alluded to Mocha Dick as a separate entity...interesting. Pandering to his ego might cause him to slip, reveal more than he might otherwise. He thoughtfully typed his reply.

Mocha Dick,

Clearly you are superior in intelligence. Help me to understand your words of wisdom. Who are the damned and where might I find you? These mystic whale files—are you their creator?

Bennett

With great will power, Bennett clicked the send button to post his article. It irritated him to no end to play the role of a sycophant. He tried to laugh it off. Too much was at stake; he would be a fool to allow his pride to take him to his grave. Ten feet under, pride was poor company.

The silence in the room, he thought too loud, and so turned on a portable radio resting on his desktop. A song from his early days of puberty crackled through the speaker with the dynamic range of a tin can, a range that only an AM station could deliver. He found the tune a direct conduit to his past. As he peered through it, he saw two boys talking in a pup tent setup in their granddad's backyard. Life and all its mysteries lay before them, their speculation as to its secrets, far from accurate. But they didn't know, and it didn't matter that they didn't know.

Now it did matter.

Bennett closed the newsreader and fired-up his compiler. He opened the *dory* file that contained his code. Once *dory* was operable, he would set it adrift out on the Internet. Not physically, but figuratively. *Dory* would understand the protocols required to communicate with open data stores on the Internet. The servers listening on the other end would send back information that *dory* would filter; looking for any traces of information relating to keywords Bennett would program it to find. *Dory* would operate day and night, strategically moving from one site to another and when finding success, zeroing in on linked sites that might contain related information. This was Bennett's mission for *dory*. With great concentration he began to finish the construction of *dory*, knowing that in his sea of problems, it might be the only thing keeping him afloat.

It started with the song of robins, and transitioned to an orange glow rising above the neighbor's roof—morning had arrived. Bennett pulled the curtain across the window to block the powerful light. After three attempts at a compile, and failing on all three, Bennett finally succeeded on the forth. *Dory* was built. There was no bottle of champagne handy to crush across *dory's* bow, but Bennett saluted and thought that good enough.

"Are you nuts!" Collin asked, after pulling an unlit cigar from his mouth and resting it on a pen. He straightened out the call slips on his desk and looked up at Bennett. "I'd die to have a woman like that, and you want to cool it with her? Women like her are scare as hen's teeth."

Bennett was leaning back in the chair across the desk. His hands were interlocked behind his head, supporting it. "I'm taking a break. I'm not sure I know her well enough to continue."

"Did you hear what you just said?"

"And your point is?"

"How will you ever get to know her well enough without spending time with her...that is, short of, hiring a detective or something."

Bennett returned a solemn stare.

"No...you aren't...can't be." Collin looked questioningly at Bennett. "Are you seriously contemplating hiring a detective? Why? She's the perfect woman. Can't you see that? I've heard of harebrained ideas, but this takes the—"

"I'm not going to hire a detective."

"Well you do have some sense to you," Collin said. Cigar to his mouth and turned it within his lips.

"I guess I'm thinking that some time away will actually help me know her better."

"Like I said before, and Bennett you know I love you like a brother, but this idea of yours is crazy."

"I thought about it a lot last night. I think it's the right thing to do..."

"Are you okay?" Collin asked, scrutinizing Bennett's appearance.

"Of course."

"You just don't look right."

"How am I supposed to look?"

"Don't know, but not like you do now."

Bennett grabbed the gavel from Collin's desk and mindlessly tapped it into his open palm. "Just didn't sleep well last night."

Collin pulled his glasses from his eyes. Bennett was always amazed at how different Collin looked without his glasses. He could almost pass as two different people.

"I've known you a long time Bennett," Collin said with conviction, his hands grabbing a cloth from his desk drawer and busily caressing the lenses in his glasses. "And I'm almost 99.9 percent sure there's something going on here that you're not telling me about."

Bennett sighed. "Let's not blow this all out of proportion. I mean, just because I'm saying I'm taking a break from a relationship with the most wonderful woman in the world, doesn't mean my entire life is in the shambles."

"Alright...alright, I can take a hint. But just remember I carry a lot more than just fat, I carry clout in this town. I can help you. Okay?"

"Sure. If I'm ever in any kind of trouble, I'll be coming to you."

Collin interlaced his fingers across his belly and eased into a satisfied grin.

Bennett returned the gavel to Collin's desk. "Oh, by the way, did you know Hoosier is in town?"

Collin shook his head looking puzzled. "Who's Hoosier?"

"You don't remember, do you?"

"It'd be hard to forget a name like that, but I guess I have."

"Summers at the Powerhouse, remember? She was a little younger than me." Bennett could see Collin's expression was not changing. "She was a counselor..."

It registered in Collin's face. "Oh yeah, the spindly little thing with the black hair in a pixie cut."

"Yep."

"Never liked her."

"You asked her out."

"I did?"

"You'd like her now."

"Big breasted?"

"And then some."

"You're right, I like her. When can I see her?"

Bennett shook his head. "Seriously though, she's back in town to rediscover herself."

Collin rolled his eyes. "So is she in one of those 12 step programs?"

166

"Nope. Her therapy of choice is to work a counseling job this summer at the Powerhouse."

"You're kidding, right?"

"No."

Collin feigned cracking the gavel against his skull.

Bennett shrugged. "I don't think its such a bad idea—"

"I was right, you do have major problems. But I can help this girl. Send her my way."

Bennett grabbed a fluorescent orange ball, belted in black with a vendor's insignia, from Collin's desk and pelted it at Collin's fat belly. Collin expelled the cigar onto the floor. He quickly recovered and picked it up off the floor.

"If you're gonna hurt me, that's not the place to aim." He laughed and his belly jiggled. "Never aim for the belly on a fat guy."

Bennett got up from the chair and headed for the door.

"By the way…" Collin began.

Bennett turned to face Collin, now in control of his laughter. He continued, "You may be right about Tanya."

"How so?" Bennett asked incredulously, thinking Collin was about to make a joke.

Collin cocked his head and narrowed his eyes. "She's asked Elander to lunch today."

"Elander? That fruit? Are you bulling me?"

"Serious as a heart attack. You can go ask him."

Bennett's voice dropped. "Is he going?"

"Do the words 'hell yes' answer your question?"

"What is she doing?"

"Is that a rhetorical question?" The cigar was back in Collin's mouth.

Bennett stepped outside the door, oblivious to his environment.

His eyes bleary, his mind on the verge of a jealous rage, his clammy hands clutching the wheel, Bennett turned down Sconticut Neck—ignoring the fears of the place that would normally settle in—on his way to West Island and the Powerhouse. Hoosier didn't know he was on his way, but he had no way to contact her.

At the causeway leading to West Island, Bennett breathed in the salt marsh air sailing across the road from one side of the bay to the other—the damp air instantly cooling the interior of his car. Getting through the rest of the day at work after his conversation with Collin had been difficult at best. He told himself not to watch Elander leave at noon for lunch, but did anyway. He even went to his window and watched Elander slide into Tanya's car. He felt numb and betrayed, but somehow managed to stay at work, his door shut, planted in his chair, staring at the painting "South Sea Whale Fishery" on the opposite wall and wondering how those men could stay out at sea for so long and how they kept from killing each other.

He came upon a small orange-railed bridge where tidal water passing between the bays squeezed under. A boy was perched on the railing and dove into the rushing current below. Two young women in bikinis awkwardly held cigarettes at their sides and laughed. A man in a ball cap sat hunched on a rock at the causeway side, his fishing pole planted securely between boulders.

At the West Island end of the causeway, what was at one time a candy store, now a boarded-up dilapidated gray shack, turned a sea blue and artic white, painted in the vivid colors of his mind's-eye, conjured from his childhood memory. He could almost feel a fireball candy scorch his tongue, much as he did as a child, probably more than a thousand times; at ten years old his almost permanently stained red tongue bore witness to the fact.

The causeway landed and turned uphill. Cottages lined it, interlaced by undeveloped land, thick in poison ivy and wild raspberries, such that a view from the air might present a checkerboard look. At the top of the hill, he saw a line of children marching through an acre of overgrown grasses, following a well-trodden path. A young man counselor took the lead and another took the tail. Behind the parade, the rectangular shaped—cinder block constructed—Powerhouse, with a freshly painted white exterior, rested inside a wall of large unearthed boulders.

He pulled his car into the grass out front. As he stepped from the car he heard a roar and turned to see a game of softball taking place in a large clearing to the side of the Powerhouse. It had to be the senior boys from West Island playing some team from up the Cape. He headed down the gravel drive to the west entrance of the Powerhouse—nothing more than a garage door. The smell of Elmer's glue wafted from within as he stepped on the edge of the blue painted cement floor of the building. Inside were a series of red picnic tables. Hoosier, wearing a T-shirt and shorts and sandals, was seated at the end of one of them. He looked up at the bare rafters, the same tents, extremely faded, appeared to be up there—the ones they used to take on camping trips to the back of the island. Hoosier looked up from her seat and spotted Bennett.

"Bennett, what a nice surprise!"

"Hope I didn't stop by at a bad time. I had no way to reach you."

"We're just working on our Popsicle stick art."

Bennett laughed. "Remember my encounter with the glue?"

Hoosier clapped her hands and bent over laughing. "Oh yes. My gosh, what a mess that was—all over the floor. And then the next day, we realized we didn't clean it all up and the picnic table was glued to the floor!"

"Glue and I never have gotten along well." Bennett could see remnants of paint smatterings on the blue cement floor. Some he was sure were dripped from his saturated paintbrush when he coated his plaster of Paris horse in red during an arts and crafts session as a child.

Hoosier nodded, a wide grin covering her face. "Does this bring back memories or what?"

Bennett took a deep breath. "Amazingly enough, it really does. Who the boys playing?"

"Camp Seahorn."

"Our arch enemy."

"Remember when big Manny was running for home and couldn't stop?"

Bennett busted out laughing. "He ran…he ran…he ran right into the backstop—knocked it clean over!"

"It didn't hurt him a lick!"

Bennett felt tears welling in his eyes, he was still laughing uncontrollably. He looked over at Hoosier who was laughing just as hard, leaning against the wall of the Powerhouse with her hand held against her stomach.

Bennett controlled his laughter long enough to say, "Robert pissed in his pants when he saw that backstop coming down on top of him!"

The two were flung into another uncontainable fit of laughter. By now the children had put the Popsicle sticks to the side and were looking outside at the spectacle of Bennett and Hoosier.

Hoosier caught her breath. "Timeout! Truce, no more funny stories for now. I'm about to bust my gut!"

Bennett wiped the tears streaking down his cheeks on his shirtsleeve. "Okay, okay. I can't take anymore either."

The two took deep breaths and tried to collect themselves. It took a few iterations. Each time one looked at the other the laughter re-ignited.

Hoosier was first to speak. "So to what do I owe the honor of this wonderful visit?"

Bennett dropped his smile. "Just wanted to run some things by you."

Hoosier scrunched her face. "Sounds serious…"

"…Maybe. When will you be done?"

"How about now?"

"What about your kids?"

"Hanna can watch them."

Bennett looked back into the Powerhouse at the spindly girl seated at the same picnic table where Hoosier had been seated. "Are you sure?"

"She's very capable...come on." Hoosier started walking towards the back of the Powerhouse.

Bennett remained planted at the Powerhouse entrance.

"Come on," Hoosier reiterated, waving Bennett towards her.

Bennett joined her on the merry-go-round in the back. She sat on the center pole with her legs crossed.

"Here?" Bennett asked, questioning the odd location.

Hoosier shrugged. "Why not?"

"Okay." Bennett stepped on the turning platform and leaned against one of the bars. "I've figured out how to beat you at your game."

Hoosier cupped her ear. "Excuse me. What in the world are you talking about?"

"You don't recall? When we were children, you sat in that very same place. If I called you a name, you would respond: 'I know you are, but what am I.' Or if I made you a compliment, you would respond: 'I know I am, but what are you.'"

"So I always won, right?"

"Until now..."

"So what's your answer?"

"To say to you: 'You're a handsome boy.'"

Hoosier flung her head back and laughed lightly. "I had forgotten all about that. Children can be so silly. I hope you didn't lose sleep coming up with that answer."

"I used it as a mental exercise."

"Boy, Bennett, you need a life."

"There is a point to all this..."

"...Yes, I'm waiting..."

"There's always an answer, you just have to search for it."

"Okay...so what are you telling me, or asking me, I'm not sure which."

Bennett nodded. "Here's a question. Why were my boots found at a murder crime scene?"

Hoosier wrinkled her brow. "This *is* hypothetical, right?"

Bennett rubbed his chin. "Not really."

"Whoa! What are you saying?"

"I know it's crazy. I just learned about it a few days ago."

"Did they take you in, are you on bail?"

"They brought me in for questioning. I think I have a very short amount of time to figure something out before they make a formal arrest."

Hoosier looked pale. "Is this for those serial killings?" Bennett maintained a stolid face, making a slight nod. "How can I help?"

"I need to fill you in on everything I know. Someone is framing me."

"No doubt." Hoosier crossed her arms and looked at the rusted corrugated steel surface on the merry-go-round. "Did you loan your boots to someone?"

"I wish it were that simple. They were stolen out of my car."

"Then it is possible, the thief, having no idea who you are, took the boots so they could not be traced back to him, killed someone and left them unintentionally behind."

"Yes, that could be, but there's another component to it all. I've been dealing with some weirdo on the Internet. He sent me a mystery file, not directly mind you, but it was obviously targeted for me. Since then we've volleyed communications. The file was encrypted, and initially I couldn't do much with it. Through a series of circumstances, I was able to run the file. The result was an image of one of the murdered victims—you know from the serial killings."

"Yes, of course."

"For whatever reason, this guy has chosen me out of nowhere to take the fall for him."

"He must be local if he stole your boots."

"More than likely, yes. The killings have obviously been local." So much for a Russian connection, Bennett thought, or somewhere else overseas.

"So how do we catch him?"

Bennett smiled with pleasure. "I knew I could count on you."

"Hey, what else have I got to do?"

"There's more to this nightmare."

"Oh?"

"I found some files on Tanya's machine. One helped me decipher the whale file, the others were porn images, including an image of one of the victims."

Hoosier's jaw dropped. "Do you think she's involved?"

"How could she not be?"

"Someone could have planted them on her machine?"

"Why?"

"To get back at you."

"You mean she could be included in this mess because of her relationship with me?"

"Yes...isn't it possible?"

"I thought that for awhile, but then I realized there was no way somebody could have networked those files to her over Pequod. She had only logged on once, and that was me testing the connection. I think she has another ISP and I guess she could have used it. Maybe that's how it got there."

"Could someone have broke into her house and physically put the files there?"

"I guess, but if so, they went totally undetected. She never hinted anything to me about a break-in."

Hoosier looked into Bennett's eyes. "It is possible, Bennett."

Bennett grabbed the bar he was leaning against and hoisted himself in the air, swinging his legs. "Yes, it is possible." But incredibly improbable, he thought.

"...Let's spin."

"What?"

"Let's spin this thing, we need a different perspective."

Bennett placed his legs back on the platform and walked to the edge. Grabbing a bar he began pushing the platform, at first walking, but once he had inertia in his favor, he began running.

"Good enough, now climb aboard," Hoosier said, her hair flying from the swift rotation.

Bennett jumped back on the platform. "This is gonna make me sick."

Hoosier flung her arms in the air and clapped. "I love it!"

After a few rotations, Hoosier found she was leaning too far to one side and lost her balance. She fell, but before reaching the platform, Bennett grabbed her. He held her only for a moment until she could get her feet under her, but he was amazed how comfortable he felt with that moment. They eyed each other until a powerful surge of dizziness overcame Bennett. "I've got to get off," he said, jumping to the ground and falling over into the grass.

Hoosier jumped too, but remained standing. "I guess these things really are made for kids," she said.

"Whew!" Bennett propped himself on all fours. "Feel like I've had too much to drink."

"What are we going to do?"

"First thing is to keep me from puking my guts out, second thing is let's get together and brainstorm a strategy. How about tomorrow at noon, let's meet at Fort Phoenix at the boat passage."

"I'll be there, but why so secretive?"

"I don't want the cops to know you're involved."

"It would appear it's too late for that."

Bennett collected himself and carefully rose to his feet. "What do you mean?"

She nodded her head across the ball field to a solid blue car parked in the street, inside the silhouettes of two individuals.

Bennett looked where she nodded. "Jeez, sorry Hoosier, I haven't been thinking straight lately. I probably led them here."

"We can't even be sure it's what we think it is. Don't worry about thinking straight. Between the two of us, we should be able to think clearly."

"—But this could be considered aiding and abetting—"

"Look, Bennett, my life is in the shambles, I need something to hold on to, a mission, a purpose, and I think I've just found one. Do you recall that summer, after I'd left, you didn't receive my well-I'm-home-in-Indiana letter until long after you expected it."

Yep. That was the summer the family van got rear ended by a semi. Hoosier was in the back seat. Her injuries were near fatal as he learned later, not until mid October or so. She told him in a letter that the doctor considered her survival a miracle.

"...I'm a survivor, Bennett. It's part of who I am. And you are as well."

He embraced Hoosier and once again sensed a warm feeling of comfort. "Let me know if they leave when I leave, or if they talk to you."

Hoosier grinned. "Will do."

"Goodbye." Bennett turned to walk back to his car.

"—And Bennett..."

"Yes?"

"It's gonna be okay."

The hiss and pop coming from the kitchen amplified, snapping Bennett from his thoughts and sending him to the kitchen, where he found white foam dropping down the sides of a pot, expiring to gas on the angry orange burner below. He turned the knob responsible for the burner to a lower temperature and poured the contents of a macaroni box into the pot. A foil cheese packet slid in with the macaroni. What the! He attempted to rescue the packet by grabbing an exposed corner—*ouch*! After thrusting his throbbing fingers under running water for a minute or two, he grabbed steel tongs from the utility drawer and fetched it that way.

He slit a package of Oscar Meyer hotdogs and extricated one from it. He leaned forward at the sink and pressed his face up near the window screen.

"You gentlemen like hotdogs in your macaroni and cheese?"

The shadows in the vehicle out by the curb showed no reaction other than the arm dangling a cigarette out the driver's window, retracted inside the window, and flopped back out.

Bennett brought himself upright from the sink. He sliced the hotdog into the turgid water churning the macaroni and returned to the kitchen window. A cigarette was flung into the street from the car window. He leaned his face up to the screen.

"You spending the night here?"

"*What?*"

Bennett hadn't expected a response from his taciturn onlookers, and was almost relieved to discover the reply came from behind a spire shaped Douglas fir in the neighbor's yard. "Who's there?"

"Shhh, it's me…"

Bennett squinted. "Who's me?"

"Alfred."

A thin figure rotated to the tree's four o'clock position, hugging against its supple branches. The tree in no way prevented the curious eyes in the car from spotting the figure.

"Alfred?" Bennett asked.

He watched as Alfred put his finger to his pursed lips in a manner so exaggerated, as to almost suggest he were kissing it. He directed the finger at his chest and then to Bennett.

Bennett lowered his voice. "They know you're there Alfred."

Alfred cocked his head. "Who?"

"Never mind. Come on in."

"Sneak me in through the back—"

"That's not…" Bennett paused, thinking what's the point in saying it wasn't necessary. "Okay."

Alfred slid in through the sliding glass door on the back porch. He immediately gathered in the smell from the kitchen with two elaborate sniffs from his beaked nose. "Macaroni?"

Bennett nodded. "Join me?"

Alfred shook his head. "Nah, the Misses put a nice plate of beans and franks in front of me earlier." He sniffed again. "Does smell mighty tasty…"

"Sure you won't join me?"

Alfred swished his lips. "She provided mighty small portions…"

Bennett started back to the kitchen. "I've got plenty for two, let's eat."

The two sat in silence at the table. The sliced sections of hotdog rolled beneath Alfred's finger from within the macaroni to a designated spot he cleared on his plate. "Hope you aren't offended. I'm a purist, not much for mixing stuff."

Bennett shook his head vacantly, mesmerized by the activity.

"I'm glad to see you're still seeing daylight," Alfred said, popping a slice of hotdog in his mouth.

"Huh?" Bennett took a sip of his tea; his eyes still fixated on Alfred's plate.

"You know...not in prison."

Bennett sat back in his chair and ran his fingers roughly through his disheveled hair. He left a crest centered on his head. "It's a mess..."

Alfred flicked the last hotdog slice in his mouth and started on the macaroni and cheese. "Want to talk about it?"

"I don't think it's best to at the moment."

"I might be able to help you out. I've read every one of the Sherlock Holmes mysteries. They say Conan Doyle had one of the most brilliant detective minds ever in all the history of man. Now that speaks something." Alfred gulped from his glass of tea and lowered his voice and squinted. "I am well versed in applying his detective principles. I can help you, son."

"This is one even Sherlock couldn't solve—"

"Such a case never existed."

Bennett shrugged.

"Now what is it you've gotten into...drugs?"

Bennett shook his head.

"...Homosexuality?"

Bennett shook his head, leaning it on his hand with his elbow resting on the table.

"That's it, isn't it? I thought there was something queer—"

"No, that's not it, Alfred," Bennett said, holding his arms out straight, beseeching Alfred to give it up.

"Look, son, in times such as these, you need something to pull you up by your bootstraps, and I can help..." Alfred's voice trailed off as he observed Bennett's face turn stern in reaction to something he just said.

Bennett began tapping his fingers on the tabletop. He furrowed his brow and asked, "Why did you want to borrow my boots?"

"What?"

"My boots—the other day—you wanted to borrow them, remember?"

"I suppose. It's not something I would make a point of remembering. After all, when you get to be my age, you're happy if you can just remember to take a BM—"

"Why would you want my boots?"

"I don't know...to go fishing? I love to spin cast, but got no boots."

"Who are you, really?"

Alfred decisively shoveled the remaining macaroni on his plate, into his mouth. "Are you okay, son?"

"You took them out of my car, didn't you?"

"Took what?"

"My boots!"

"Son, you're scaring me—"

"You're not telling me the truth about yourself."

"I swear..." Alfred slid his chair back and stood up. "I should be going."

Bennett did the same, shifting into Alfred's direct path to the sliding glass door. "Why are you framing me?"

Alfred's lower jaw began to tremble. Beads of perspiration oozed on his shiny forehead. "Son, I think you've gone mad—"

"Mad? I've gone mad? You're the one who's mad. Only a mad man could have killed—" Bennett stopped himself.

Alfred cocked his head. "Please, get out of my way. The Misses will wonder where I am. I've got to get back to home."

Bennett pointed his finger at Alfred's chest. "You're not leaving until you've told me the truth."

Alfred's eyes darted about the room. He moved to his right, Bennett followed suit, blocking him. He moved to his left, Bennett again blocked him. "Blahhh!" he yelled, and sprang at Bennett. Bennett jumped back in surprise. Alfred flung himself over to the sink and yelled through the window screen. "Help! He's trying to—"

Bennett, with great speed, cupped his hand over Alfred's mouth and muffled the cry midstream. "Alright, you can go. Just stop screaming. You promise to stop screaming?"

Alfred nodded his head as best he could.

Bennett could not risk him stepping outside and talking directly to the cops. He took a deep breath. He had to turn this confrontation around. "Sorry, Alfred." He took a deep breath. "I've been under a lot of pressure as of late. I overreacted. I consider you a friend. Please...forgive me."

Alfred nodded again. Bennett slowly removed his hand. A quick glance out the window told him the car was now empty—the doorbell rang. Alfred straightened his clothing and clicked his heels. He stared into Bennett's eyes and nodded. "I believe you. I'll take care of this."

Alfred opened the front door to find the cops standing on the porch. One of them gruffly asked to come in. Alfred nodded and the two entered. Bennett remained in the kitchen. A few moments later Alfred reappeared. "They've left," he said.

"Thanks, Alfred. What did you tell them?"

"Loud television."

Bennett knitted his brow. "The television isn't even on…"

Alfred tapped his temple with his finger and winked. "Leave it to Sherlock, my son, leave it to Sherlock."

From his home office window on the second floor, Bennett looked into the street at the pea green car parked beneath the neighbor's mimosa. A shift change: the previous car had been a sickly gold color, not unlike the dinner he had just tossed in the toilet. He had heard of people vomiting as a result of being extremely nervous, but had never believed that sort of thing really happened—until now. He contemplated his situation: a prisoner in his own house, the Gestapo watching the yard perimeter, the death penalty…he paused, his stomach mounting another scorching assault on his already sore throat. Deep breaths; take deep breaths. Man, I attacked Alfred! I've got to get a handle on myself. Think. Whew, boy. He scanned his office, stopping at the CD tower on the floor. Leafing through the jewel cases, he found what had made him stop there in the first place: *Debussy's Clair de Lune*. Under normal circumstances classical was out of the question, better for an afternoon nap then a march through the web, but these were unusual conditions and the strumming of his nerves with a dulcet instrumental sounded as good as a glass of ice water would someone stranded in the desert. The music started and so did the PC.

What have you found Dory? He viewed Dory's output:

Matches found with high probability:	128
Matches found with medium probability:	1796
Matches found with low probability:	5893

Bennett's heart pounded when he saw the numbers—voluminous—probably 99 percent worthless junk, but maybe, just maybe…He decided to review the high probability matches. He began a meticulous march down the report Dory had built for him. In the first column was the match with a limited amount of surrounding text to provide some means of context, and in the second column, a reference pointing back to where Dory found the information.

…do not feel White Whale is safe in terms of…	http:www.kr>
…the problem with White Whale is it's inabi…	nntp:alt.bu>
…I would not recommend White Whale Inc. for ma…	nntp:alt.bu>
…the fact is, white whale destroyed the peace…	pop:bthom@z>
…got so blasted at the White Whale last nig…	pop:jsmitt@>
…Melville's Moby Dick is without a doubt…	http:www.fj>

Bennett blushed as he scrolled the list. The *pop* entries embarrassed him greatly: they were captured from private mailboxes. Amazing that Dory could get them in the first place, even minimal security would have prevented it. He didn't relish the idea of reading private e-mail, but he weighed the sleazy behavior against his heightened level of desperation, and concluded he was justified. No entries for Mocha Dick. Moby Dick and White Whale made up all the entries. After scrolling all 128 listings in Dory's report, Bennett began viewing each one. He would start with the e-mails first. For the information found in the e-mails, he crafted an ambiguous message to each potential recipient.

"Could you please help me? I'm contemplating working with the White Whale. A friend told me you've had dealings with the White Whale (If not, please ignore this message). Any information you can provide me that might aid me in my decision would be greatly appreciated.

P.S. The name could be White Whale, Moby Dick, or Mocha Dick, I've seen all ways.

Best regards,
Frank Gerald"

Yeah, and this is gonna work? It was almost laughable.

Bennett went out on the Internet and located a site that provided free web-based email accounts. He spoofed his machine's Internet address, not sure if the site would capture it. He filled out the request form with bogus information that could by no means or measure be traced back to him.

An hour and a half later, Bennett had visited all the web sites and newsgroups, without learning anything other than there were a lot of businesses using the name White Whale. He wrote a quick script to send out his canned email to all the accounts Dory found with matches, using his bogus email account as a return address. He looked out the window; now dark outside, only the reflection of the streetlight on the car's front chrome bumper gave away its existence. He shut the window and pulled the curtain. The CD had played through twice—time to shut it off. He headed downstairs, microwaved a cold cup of coffee and shivered; the cold deep inside still prevailed. Thunder broke at a distance and tiny pellets of rain pattered against the windowpane above the sink. He pulled the window shut.

The rain pummeled the windshield as Bennett drove cautiously down 6th Street on his way to Pequod. He cursed the rains, which now had settled comfortably over southern Massachusetts. Betrayed by sleep, his eyes felt as sodden as his shoes after their plunge in the water current tumbling across his sidewalk leading to the garage. He came to a halt at Market Street where traffic stopped. He took a kleenex from the glove compartment and wiped the windshield to remove a foggy coating on the interior of the glass. Through the clearing he saw down the road a crowd of people moving about on a corner. Had to be a wreck. He thought it must have just happened, as there were no emergency vehicles or police in sight. He turned on the radio, heard a talk show discussing the murders, and quickly turned it off. Doesn't this town have anything better to do? What happened to all the talk about the Red Sox? Traffic inched forward.

A cloudburst churned through the trees by the time Bennett reached William Street. His windshield was re-coated with opaque condensation. Once again, he reached forward to clear a section of his windshield. He did so, just enough, to see the car in front of him. He moved a few hundred feet and stopped. Between the wild wisps of the wind, he heard voices, voices shouting a muffled phrase. He wiped more of his windshield. Now through the bigger section he watched people in bright red and yellow foul weather gear march in front of City Hall. They carried signs high above their heads with writing in bright colors streaming down the cardboard. The wet and distorted letters looked eerily gothic, but readable:

"The murders stop now!"

"Catch the creep!"

"I vote special dispensation!"

Their chanting slipped through the tiny crack Bennett opened through his window.

"Death to thee..."

The signs circled one after another before him.

He gasped when he thought he could make out his name.

The bearer followed the circle and turned his back to him. Bennett panicked, not sure what he had read. He followed the sign to the backside of the circle. The car in front of Bennett leapt forward and seconds later the car behind began to honk.

The sign bearer was about to turn the corner and face Bennett once again.

The car behind honked three times.

Just one more second...

He squinted as the sign made the corner and was flashed before him. The letters were streaking worse than before and he couldn't be certain what they stated.

Bending over to hide his face, he wiped the sweat from his forehead, hung a tire-screeching left, and sped down Mechanics Street, leaving the cries behind. He thought: a driving rain, strong winds…marching in it?—the people of New Bedford were serious—they wanted their man captured—captured now. Bennett swallowed hard. Could it get any worse?

"Hey buddy," Collin said, as Bennett entered the reception area at Pequod. "You've got a visitor."

Bennett removed his jacket and shook rivulets from it. "Who?"

"Some guy from the FBI." Collin looked at Bennett oddly. "Are you in some kind of trouble?"

Bennett scratched his neck, buying time for an answer. "Just a little work on the side."

"The side?"

"Yeah, you know, hacker stuff."

Collin's face drained white. "You're not going after our friend are you? The whale file guy?"

"Well, as a matter of fact—"

"Well holy cow, Bennett! What gives you the right to make that kind of decision without consulting the rest of us? I thought this was your pet project, something you were going to keep to yourself."

Collin looked furious. The veins in his neck protruded, feeding blood to his face, coloring it. Bennett stepped back from him, half expecting a violent outburst that involved physical contact. "Look, Collin, I had to—"

Collin pointed his finger at Bennett's chest. "You didn't have to do anything. You chose to do this."

Bennett took a deep breath and sucked in as much air as he could manage. "I don't get it—what's the big deal?"

"The big deal is you made this decision solo—without even considering the opinion of the rest of us."

"Look," Bennett said, holding out his hands, "I had to okay. You need to take that on faith for now."

Collin shook his head and managed to relax his voice. "I don't see how Bennett…", he continued shaking his head, "I don't see how. And I don't know why you carry on with this…"

Bennett watched as Collin turned his back on him and walked decisively down the hall to his office. All the years he had known him, Collin had never expressed such anger directed towards him. He knew he had a temper, he had seen that, mostly directed at drivers and waiters, but he never thought he would find himself at the other end of the tirade. He thought he might have to tell Collin what was going on. He wondered why it was he was hesitant to do so in the first place. Maybe he felt embarrassed or ashamed—thinking others might believe it all true, that he was indeed the killer.

The aroma of coffee, steaming from the cup Bennett shakily poured, didn't elicit its customary sense of pleasure. He stood there, in the kitchen, gingerly sipping from the cup and staring at the brown speckled tiles on the floor. The brown spots were circular, variegated, and at times converging. He heard laughing down the hall. Chanting voices echoed past him, obviously from Elander. What a piece of work! He became aware of a heightened sense of hearing—amplified sounds echoing and skewing, warping into the realm of the surreal. A drop of coffee splattered near his shoe, catching his attention. Settle yourself. He wondered how that could have happened and then noted how vigorously his hand quivered, generating black combers in his coffee. He set the cup on the counter, eyeing his reflection in a chrome paper towel dispenser as he did so. Could he be seeing himself? The face looked haggard and bland, every ounce of character spent. Need to get some sun or something—a loud crash caused him to jerk. His eyes jarred wide, he instinctively jumped back, observing the coffee cup smashed on the floor, the dark contents streaming tentacles under the refrigerator.

"You set your cup on the edge."

Bennett whirled around, facing DeWald. He tried to recover gracefully. "Wasn't paying attention. You know how that happens. Mind wondering an all…"

DeWald nodded. "Perfectly understandable." He bent over and surgically removed the broken ceramic pieces from the cup, placing them in a trash dispenser. Bennett watched, wondering why DeWald was helping. DeWald then floated a handful of paper towels from the dispenser onto the remaining puddle. After a few moments of allowing the towels to soak up the liquid, he gathered the towels up and added them to the trash. Moments later the floor was dry.

Bennett awkwardly looked on, wondering why he didn't make a move to cleanup his own mess. "Thanks…" was the best he could muster.

DeWald wrinkled the corner of his mouth and cocked his head in the direction of the hallway. "Let's go to your office."

Bennett followed DeWald to his own office. DeWald motioned to Bennett to have a seat in the chair behind the desk. DeWald sat across from him. He pulled a notepad and pen from his coat pocket. "Take some deep breaths, Bennett."

Bennett scratched his neck. "I'll get around to it sometime, right now I'm not sure I have time to breathe. You here to arrest me?"

DeWald shook his head, not removing his eyes from the notes scribbled on the notepad. He looked up. "Actually I'm here to help you."

"Why do I have a difficult time believing that?"

DeWald raised his eyebrows and shrugged. "I don't know, you tell me." He cocked his head and scratched above his ear. "You know, Bennett, if you've got any chance at all...any at all, it's sitting here in this chair. What have I got to do to convince you I'm not putting on some act to get a confession out of you—how stupid do you think I am? Obviously you're a highly intelligent individual. I'd be wasting my time trying to pull one over on you. You know it and I know it. So lets drop our preconceptions and get busy—time is critical and we're wasting it discussing something that should not even be a discussion point." DeWald dropped his voice. "Let me help you."

Bennett took his deep breaths. "Answer one question..."

DeWald bobbed his head from side to side as if contemplating his answer, but finally said, "Okay."

"Why?"

"Why help you?"

"Yes."

"Because in all honesty, I believe you're innocent. I believe you've been framed."

"Is that just a gut feeling or a theory based on facts?"

"Does it matter?"

"I'll feel better, if it's based on facts. I'm a facts type of person."

DeWald sighed. "Well then, you won't be feeling better, but..."

Bennett rolled his eyes.

"...I've been a cop for ten years and I can read a suspect. I don't think at this point in my career I can be fooled."

"At least one of the cops in that room—that older fat guy—has been a cop obviously longer than you have, and I *know* he thinks I'm guilty. I think he's about ready to place an order for some rope—you know, the thick kind, the kind that—I'm talking www.hangingropes—"

DeWald held up his hand. "I know the kind." He tapped the notepad with his pen. "In his case, the pressure to solve a crime can have that affect."

Bennett, ever so slightly, nodded.

"Now," DeWald continued, "let's have a time of sharing, shall we?"

Bennett turned to face his PC and brought up a blank document in his word processor. "Okay, what can you tell me?"

"Actually..." DeWald began, "I thought we'd start with you."

Bennett sighed and removed his fingers from the keyboard. "What do you want to know?"

"Well what do you know, the rain's stopped," Patrick said, leaning against the glass window in the front of the bar and peering through the painted pig's eye.

"There went your excuse for not going through with your plans," Karl said, standing behind the bar, polishing the shot glass resting in his hand with a bar towel.

"I don't know, doesn't seem to interest me much at the moment, you know."

Karl put down the glass and picked up another. "You need a break—need time to grieve over Paul's death. I'll take care of things."

Patrick moved away from the window and looked at his fishing tackle stacked in the corner of the room. "We could play cribbage—"

"Will you just leave."

"Can't take another lurch, huh?" He grabbed the tackle from the corner and carried it out the front door, the bright sun temporarily illuminating the cave-like bar.

The tips of the two fishing poles in the back seat of Patrick's car jutted out the rear window. Two Rebels, one with a bright blue back, the other a gray, clinked their tri-hooks against one another. The marina was just a few miles down the road and Patrick was there moments after he left. He grabbed the rods from the back seat and his tackle box from the floorboard and walked to the pier, the rods in one hand, the tackle box in the other. As he approached the *Where II* he noticed a man leaning against a wooden pylon next to his boat's quay, the man's arms folded across his chest and a cigarette burning in his mouth. It was the detective.

"Don't you know not to smoke around gas powered boats?" Patrick said, ignoring Thomas and placing the rods and tackle box in the stern of the Whaler.

Thomas removed the cigarette from his mouth and squatted on the quay, looking down at Patrick. "Interesting boat you have there." He blew smoke in Patrick's direction.

Patrick began removing rope from the stern cleat on the quay. "What's so interesting about it?"

"You in a hurry or something?"

Patrick looked up into the sun at Thomas, his face glistening with sweat. "Not especially, would like to go fishing some day though."

"Fishing, huh."

Patrick ignored him and began moving towards the bow to remove rope from the cleat there.

"Interesting setup your boat has," Thomas said.

"What do you mean?" Patrick began untying his boat from the bow cleat.

"Hold on a minute there, I'm not out squatting on this dock for my health."

Patrick sighed. "Obviously not, with that cancer stick in your hand."

"I want you to explain a few things to me."

"Such as?"

Thomas pointed towards the floorboards at the stern of the boat. "Like why you have a hidden compartment there."

Patrick furrowed his brow. "If you say so." He pulled the boat key from his pants pocket. "So what, did you search my boat?"

Thomas gave a lethargic nod and slowly jutted his jaw from side to side. "Just taking a casual stroll along the dock here and happened to look in your boat. That loose board there, the one hiding the secret compartment down under it—it's in plain view…"

"So I've got a loose floorboard. Is that a crime?"

"You sneaking in drugs?"

Patrick laughed snidely. "Give me a break, officer," he said, fully aware Thomas was a detective.

"Or, maybe you're hiding something else underneath there."

"You'll never know." Patrick chewed his tongue.

"Pretty bold statement from a man who's a search warrant away from being busted."

"For what?"

"That blood," Thomas said, directing his head with a nod towards the steering column, "there at the base—fish blood? Maybe, could be something else, I suppose. A sample provided to one of our fine labs up in Boston might reveal some interesting results. What do you think Mr. Arnold?"

"So what—your boys up in lab are going to have an orgasm over bluefish blood?"

Thomas caressed his lower lip with his tongue, ending with a smacking of his lips. "Bluefish blood…bluefish blood…" He burst out in a laugh.

Patrick freed the rope from the bow cleat on the dock. "Glad I could entertain. See you later, officer," he said as his lonely middle finger caressed his jaw. The *Where II* idled away from the quay and once beyond the controlled speed perimeter, Patrick viciously threw the throttle forward, the Evinrudes reacting with a guttural scream.

"Provide me an explanation for all the circumstances surrounding your life that makes you the perfect suspect," DeWald asked, licking his index finger and shifting pages in his notebook.

Bennett sank back in his office chair. "An example being?"

"All the whaling paraphernalia in your home."

"That's a crime?"

DeWald sighed heavily. "Look Bennett, you're gonna have to stop being so defensive. Do you realize that if this thing goes to trial your entire private life will be on public display? It's the old saying: you can tell me now, or tell them—the public—later."

Bennett frowned and shrugged. "I don't know...I guess it's something I grew up with—my dad kept the stuff around, in fact most of it's his."

"Helps you remember him?"

Bennett looked at DeWald inquisitively. "How did you know my dad was dead?"

"I'm a detective, I investigate. Right?"

"I'd like to leave my dad out of this as much as possible."

DeWald nodded amenably. "Okay, let's just leave it that you keep it as a fond remembrance. But you belong to the NBWHS too?"

"Yeah, me and almost everybody I work with. We're a clique, we get into the whaling stuff."

"Yes, but you must agree it's in an odd sort of way?"

"The whaling stuff? How so?"

"Well, most people involved with whales are into saving them, you know...the tuna nets and all."

"That's dolphins. And we're for saving whales too. We just happen to find the history of whaling alluring."

"It's like flying a Rebel flag."

Bennett gritted his teeth. "Damn, are you Kathy Rueger, or a detective?"

"Interesting you should bring up her name with such emotion," DeWald said, reading his face.

Bennett shook his head and stopped suddenly, wrinkling his forehead. "Wait a minute. Why are you asking me about this whaling stuff? What exactly am I suspected of doing?"

DeWald leaned his head back.

Bennett turned his head to the side, his eyes locked on DeWald. "There's a connection, isn't there?"

"I can't say," DeWald said, attempting a poker face.

"You wouldn't be asking if there wasn't."

"Could be I'm trying to establish a motive for the Rueger murder— she highly offended your group, and your group, after all, is heavily involved with historical whaling."

Bennett grinned. "Yes, I guess that could be…"

"Look Bennett, if you read between the lines during this interview and draw some conclusions of your own, then what can I say? Or, for that matter, do about it?"

Bennett rubbed his lower lip with his index finger and stared absently at the floor. "So there is a connection…"

"I don't know what you're talking about, Bennett."

DeWald asked Bennett about the boots and Bennett surprised himself by wasting no time in implicating Alfred. DeWald followed with standard questions concerning where Bennett was on certain days at certain times and could anybody vouch for him being there. He asked who might have a motive for framing him. Bennett could only shake his head. DeWald closed his notepad and shoved it in his coat pocket. "I'll be in touch with you, Bennett."

"Wait."

"What?"

"My computer, what did you find on it?"

"What should we have found on it?"

"This is not Jeopardy."

"I can't tell you that. I'm sorry."

Bennett cleared his throat and averted his eyes. "If you were to have found some images on it, say some unusual images, they aren't mine. I got those from someone else's—" Bennett stopped mid sentence.

DeWald leaned forward. "What were you going to say? Are you protecting somebody?"

Bennett sat in silence.

"They wouldn't do the same for you, Bennett. Are you willing to die for this person, and a guilty person at that?"

"I don't know…I mean I don't know, *yet*. I can't decide—"

"Is it Patrick?"

Bennett was taken aback. "Patrick? How did his name come up? He doesn't have anything to do with this."

DeWald nodded. "Okay, if you say so…"

"What are you trying to say?"

"I'm not trying to say anything. I'm just thinking about who's close to you—who you would try to protect—even if it meant losing your life."

"You'll have to look further—he and I are not close."

DeWald flipped his hands out. "Okay, if you want to be a fall guy, then so be it."

Bennett shrugged and pushed his fingers through his hair, still damp from the morning rain. "Look...I want to discuss something you may not be aware of."

DeWald looked at Bennett with interest. "Go on."

"I've been in an electronic wrestling match with this guy."

"Electronic?"

"The Internet. He's been sending me encrypted files and then taunting me in his emails for failing to decrypt them."

"Who is he?"

"No idea."

"What's his handle?"

"Handle? He's not using some damn CB."

"You know what I mean."

"He goes by Mocha Dick, but has alluded to himself as being Ahab."

DeWald nodded, consciously maintaining his facial expression. "Split personality. What were the names of these files?"

"He named them *whale*."

"Help me get a copy to the lab and we'll see if they can't crack it."

"I've already cracked one."

DeWald raised his eyebrows. "And..."

"It generated a file named *ambergris*."

"*Ambergris*?"

"The name is not important, you can look it up. But what is important is that it was an image file and it showed what looked like a dead girl. She had a whale tattoo on her forehead. I printed it out, I'm sure you guys found it." Bennett noted DeWald's surprised expression. "Why do you act so surprised? Both these files were on my machine."

DeWald lost his look of intrigue. "Probably..." he said, and carefully noted Bennett's words on his notepad. "If you don't mind me asking, how did you decrypt the file?"

DeWald's expression and answer made Bennett contemplate the issue that maybe the files were not on his machine, that maybe somewhere on his machine roamed a malignant process, spawned by his actions to decrypt the file, that was into self-cleaning, sort of picking up after itself, leaving no tracks. Anything was possible. He had to think for a moment, but he recalled the question DeWald had just asked. "I found another file named *harpoon*. If properly typed on the command line, it will decrypt the *whale* file and generate the *ambergris* image."

"What do you mean by 'properly typed'?"

"I don't recall if it was a pipe, an argument, or a redirect. But a combination thereof will work."

"I don't know what you just said, but you can do it again, right?"

Bennett nodded and added, "The guy is trying to emulate his murderous ways in cyberspace, isn't he?"

"Can't answer that one."

Bennett's eyes lit up. "If that's the case, the victim *is* the *ambergris*—are you finding victims in the dead whales?"

DeWald repeated himself.

Bennett's mouth opened, cheerfully aghast. "Perfume is feminine—yes, this is starting to make some sense!"

The *Where II* rested on the beach one half mile down from the marina where it docked. Footprints in the sand ran parallel to the beach in the direction of the marina to within one hundred yards or so, before veering sharply up towards the parking lot. Over the horizon came Patrick, a white 10-gallon paint bucket with its thin metal handle clutched in his hand. He followed the footprints back to his boat. He threw the bucket onto the bow, and when it hit, it bounced, tossing a brush on to the deck. Patrick leaned into the bow; his feet planted in the sand, and pushed the Whaler back into the water. He waded out to waist level and pulled himself up the ladder he had placed on the side of the boat.

As he took the controls and gunned it, he shuddered, hoping Thomas had left and not been watching from the shore. After his encounter with him at the dock, he had headed out through the New Bedford dike to open seas. After 30 minutes, he turned around, thinking Thomas would have left by then. He needed to pick up some materials to clean the boat.

The more he thought through his clandestine actions, the more defensive he became. What of Thomas seeing him? Every man is entitled to clean his boat, especially with the mess a lively flopping bluefish can leave on a boat.

At a mile or so offshore, he lowered the Evinrudes to trolling speed. He scrubbed every inch of deck space with soap he added to seawater. He placed the rods in the holders at either corner at the stern and let out a hundred or so yards of line. It took two passes, but on the second one, the tip of the pole on the left suddenly swung from its erect position to one where it paralleled the water. Patrick pulled the rod from the holder and began reeling. It fought with great vigor—had to be a large blue. He watched his line cut through the water, first in one direction, then in the opposite. Moments later he gaffed the fish from the water and watched it fly into the stern. It was a ten-pound blue, maybe more. He lifted the lid on the starboard under-gunnel rod rack and pulled out a fillet knife. The belly of the blue was slashed and it was left to thrash about the deck of the boat, the tail madly drumming the hollow transom, clumps of blood and internal organs smattering the deck and sides of the boat.

Patrick stood back and watched as the thumping decreased. Sorry, my friend...

Chapter X

A smashed pack of cigarettes, its loose plastic case fluttering, skidded past Hoosier on the smooth blacktop surface banding the top of the New Bedford hurricane dike. Her eyes locked on the pack for less then a second, returning cautiously to things further down the road. The road sprung from Fort Phoenix and traversed the protective hurricane dike that spanned from Fort Phoenix, across New Bedford harbor, to New Bedford proper. The blacktop surface of the road shimmered in the intense sun that earlier had remained cloaked behind the morning clouds and skirts of rain. A pair of bikers whizzed by so quickly that Hoosier felt her shorts lift. She looked down the road, to the small bridge in the middle of the dike, the one passage for boats. Seagulls were active on the large, near flesh colored stones that made up the skin of the dike. Running on motor, a tall-sail ship gracefully approached the passage through the dike.

Hoosier casually looked over her shoulder. No one appeared to be following her. She even looked over the rocks and the bay, for all she knew they could be anywhere. The dike stopped at the bridge, not really a bridge at all—at least not one someone could cross and picked up the dike again on the other side. She leaned over the railing and watched the swirling green water below, the tall sail gliding through as she did so, the top of the mast just beyond her reach.

"We've got to stop meeting like this."

Hoosier turned to see Bennett leaning on the railing next to her. "Where did you come from? I didn't hear anything."

"Ninja training…" Bennett winked. "Class of '72."

Hoosier's short brunette hair shined red at the crest in the strong sun. Bennett could see himself smiling in her mirror-lenses resting in wire frames on a delicate nose. She smiled back, dimples accentuating the warmth of her expression—the same dimples she wore many years ago with her pink ballerina outfit in the Polaroid Bennett still possessed, and looked at each time he cleaned his closet—invariably stumbling across it in a photo album on the top shelf—she stood with him in front of a large mineral rich rock that occupied space in the front yard of her cottage, her head below its top, Bennett's just above, and he dressed as a diver with nothing more to convey this than a diving mask and snorkel. The image had stuck with him and he could still place that ballerina's face on Hoosier's.

"Hanging in there?" Hoosier asked, placing her hand on Bennett's shoulder.

Bennett bobbed his head as if mentally weighing good and bad in the balance. "Yeah. A little rough at times, but…yeah. Had a talk with Detective DeWald this morning."

The grin left Hoosier's face. "Oh…"

"No, no, it was good, very informative actually. It seems the real killer is into killing whales also."

"You mean whoever has been killing the whales has been killing the young women?"

"I'm almost sure of it. Which means this guy I've been dealing with on the Internet—Mocha Dick—is probably the killer. He's killing whales and placing his victims inside."

Hoosier gasped. "How sick!"

"Well I don't know this for certain. But DeWald didn't outright deny it."

Hoosier involuntarily shivered. "Oh, that is so sick. Those poor women."

"Well my guess is that they're dead by the time he places them in the whales."

"Gross! That is just too weird. What a loon!"

"He's probably using the Internet to recreate his crimes—to stir his imagination to relive them, or maybe to communicate them to others. He's probably a real eccentric of some sort. He'd almost have to be to use command line executables to operate his emulation of his real world crimes."

"What's so eccentric about that?"

"Fifteen or even ten years ago, it made sense, but today with virtual reality, a very graphic intensive technology, why not use that? If he's trying to virtually relive his crimes, why not use the most visually-aware technology, not boring and bland command line programs."

"Maybe he doesn't know how to use that technology."

"No, he's too sharp. The programs may be visually deprived, but the inner workings are pure genius: an ability to pass through firewalls, manipulation of network services, exploitation of esoteric features found only in the far corners of operating systems, hijacking benign protocols for malicious penetrations. And on and on…"

"I'm not sure what you just said, but do I detect a smidgen of respect?"

"For the skills, yeah."

"Keep in mind the imagination is much more powerful than graphical images. Just compare the experience of a reading a book with watching a movie."

"Touché."

"So what do we do?"

Bennett sighed. "The guy lives in two realms…" He paused.

"Yeah…" Hoosier said, encouraging him to continue.

"I'm no damn good at catching this guy in the real world. I don't see myself knocking on doors, asking people questions, visiting the crime scenes—"

"I highly recommend against that. They expect a killer to visit crime scenes."

"You've been watching too much of the Discovery channel." Bennett grinned. "But…I do see myself as damn good at catching him on my turf."

"The Internet?"

"Yep. If I catch him there, I've caught him here."

"How do I help?"

"We need to devise a strategy. The question of the day is: how do I track this guy down in cyberspace?"

Hoosier wrinkled her brow and then broke into a smile. "To quote Mary Howitt: 'will you walk into my parlor? Said a spider to a fly; 'Tis the prettiest little parlor that you ever did spy…'."

Bennett returned a blank stare before his eyes opened knowingly. "You're a genius!"

"We just have to come up with the right bait."

"Something his insatiable ego can't resist, something to make him drool."

"What has drawn him so far?"

A powerboat ripped through the canal out towards the open seas. "That looks like Patrick's Whaler," Bennett said, his eyes following the boat as it slammed into the waves of the incoming tide, water splaying at the sides of the bow while the engines groaned at each collision.

"Could it be?"

"Could be, it sure looks like his boat, it even looks like him from the back." Bennett shook his head. "I'm sorry, where were we?"

192

"Oh, I was asking what would bait him in?"

Bennett pondered the question thoughtfully. "He likes to parade his computer prowess."

"So taunt him, challenge him. Lure him in someway."

"From where though?"

Hoosier pursed her lips. "I don't know, how about—"

"I've got it—a web site. I'll set up a web site. I'll use it as my parlor. What better parlor then a web?"

"With what content?"

Bennett caressed his chin. "A serial killers' showcase."

"But why will that attract him?"

Bennett thought for a moment. He mentally reviewed the email he had received from the man. "I'll put something up he can't resist. I'll have images of all the famous serial killers and then one that is blank with a question mark. Underneath I'll say something to the effect that rumor has it there's a cyber serial killer out there, the first of his kind—really blow it up—I don't think he can resist that."

Hoosier nodded. "You may be on to something."

Bennett grinned. A couple of bikers leaned their bikes against the concrete roadside and walked up to the railing. They looked almost alien in their helmets and stretch pants.

Hoosier dropped her voice. "I hate that we would play a role in glorifying those sickos."

Bennett wrinkled a corner of his mouth. "Yeah, I know what you're saying. Unfortunately in order to catch slime, you have to enter slime."

Hoosier grinned. "Just as long as we can wash it off when we're through."

Bennett paused. He looked down into the green and blue waters below, watching the water rush against the piles and reseeding to expose communities of barnacles. Why did he choose me? Why pull me into this? "I guess he's framing me. Playing this game thing with me on the Internet is not enough. He wants me put away. And put me away for good. Whoever this is must really, really hate me."

Hoosier scrunched her face. "But how could you possibly have someone hate you this much and you not know it?"

Bennett nodded and smiled. He turned his head from the crashing waters below and looked into Hoosier's wind blown face. "That's the 10,000 dollar question."

On the following day, from the time he received the phone call at work, to the time it took him to walk to his car, the fact that Patrick called and asked to meet him at the Uncas, still had not adequately registered with Bennett. Patrick never called, ever, until today. His conversation had been short and to the point. Bennett arrived at the car and changed his mind on driving. The day was turning out pleasant in an unusual sort of way: overcast but dry, a cool, light breeze from the bay, a nice detour from the brilliant sun of the day before.

The smell of crab cakes and lobster rolls commingled around Frank's outdoor café where the patrons seated themselves at the white round patio tables with alternating yellow and white umbrellas. As Bennett strolled by he spotted a lone man reading a newspaper. The man looked up as he passed—a cop watching? Bennett could add paranoia to his list of recently acquired mental illnesses. A block down the sidewalk he took a glance over his shoulder and sighed, finding the man remaining in his seat and dutifully studying the paper.

The eye shock usually encountered when entering the Uncas, when going from bright light to almost complete darkness, was greatly abated by the overcast skies. Bennett gave his eyes a few moments to adjust, and began scouting the bar for Patrick. He hadn't seen Patrick's car out front, and so did not expect to see him in the bar, but to his disbelief he found him, huddled in a booth and girdled in a twenty-year-old pinstripe suit.

"A wedding?" Bennett asked, slipping into the booth seat across from Patrick. The smell of stale beer rose from the cushion as it expelled air from his weight. Patrick had a fresh one in a tall glass resting on the table in front of him, next to a pile of discarded peanut shells. Tiny streams of beads slid down the side of his glass. Bennett grabbed an unshelled peanut from the plastic basket on the table and cracked it open, shooting the contents into his mouth.

Patrick took a shallow sip from his beer. "Funeral. They found Paul's body two days ago. At least what was left of it."

Bennett nodded, not saying anything. His words would ring hollow and he knew it, so why bother. But in a strange way, knowing that Paul was dead did sadden him. Patrick would never believe it though.

"After the Coroner gave him the once over and determined his death was accidental drowning, his mother had him cremated. She's a Christian Scientist. They always cremate. I wanted to say goodbye. I felt moronic talking to an urn. It was me and Paul's mother, and that's all that showed up. Nobody from school, none of his old girlfriends. You know he was a idiot, knew how to get on my nerves like nobody else could, but when I saw that urn and realized that was all that was left of him, it kind of shook me up."

Bennett thought he detected tears welling in Patrick's eyes, but he couldn't know for sure in the darkness of the pub. Patrick gulped his beer until the glass went empty. He grabbed a handful of shelled peanuts and began cracking them open and gulping the peanuts inside. He made no effort to say anything.

"So where's you car? I didn't see it out front."

"It's out there. The blue Camero."

"New car?"

"Impressive, huh?"

"Impressive yes, but how did you afford it?"

Patrick shrugged. "Things have gone good at the bar, you know. Promotions and so forth."

"What? All five patrons are now drinking themselves into a stupor every night?"

"Hilarious. Here I am grieving...no class."

Bennett lifted his open palm to Patrick and nodded. "You're right. It's a cool car, congratulations."

Patrick nodded blankly and fixed his eyes on his empty beer glass, mechanically shoving more peanuts in his mouth.

"So, what did you want to talk about?" Bennett said, after realizing Patrick might sit in silence for a long time to come.

Patrick stopped chewing and brushed some shell fodder from his shirt with his hand. "Just wanted to see how you were doing."

Bennett eyed the remark with skepticism. "You want to know how I'm doing? Why are you so suddenly concerned how I'm doing?"

Patrick tilted his head. "Granddad wants us to try to get along. I thought I'd try."

"Okay," Bennett said, trying to sound like he believed it. "In that case I'm doing just fine."

"No problems?"

"What problems could I possibly have?"

Patrick talked with expectation in his eyes. "Who knows, right? Everybody has problems. I heard the cops had been around."

Bennett tried to hide his surprised reaction. Who could have told him? Was it a lucky guess? What was his motive for trying to guess? Maybe a neighbor told him. Hoosier...could it have been Hoosier? No, Hoosier wouldn't say anything. Bennett grinned. "Now where would you have heard a tale like that?"

Patrick shrugged. "I'm a bartender. I know what freakin' color of underwear the woman at the corner 7-11 wears."

"But not what I've been up to. No truth to it."

Patrick pursed his lips and nodded.

"In fact," Bennett continued, "I'd heard the cops had been around your neck of the woods..." He went fishing with the open statement. It had bothered him the other day when DeWald brought up Patrick's name when asking him who he was trying to protect. What did DeWald know about Patrick that would bring his name up first?

"The cops are always in my neck of the woods. They've built a freakin' cabin there."

"So you're not in any sort of trouble?"

"Hell no, no more than usual that is."

Bennett knew what he meant by that. The list included disorderly conduct, urinating in public, a handful of DUIs.

Patrick pulled a quarter from his pocket and flipped it into the air, catching it in his outstretched hand. "One last one for Paul."

"One last what?"

"Jukebox tune."

Patrick got up from the booth and walked in the direction of the jukebox placed at the end of the aisle. Bennett had an eerie sensation that the two of them were feeling each other out. He knew he was feeling Patrick out, and he was almost certain Patrick was doing the same to him. But why? Patrick never gave a damn before what he did. It made no sense. Perhaps Paul's death shook something loose. He watched as Patrick placed the quarter in the jukebox and listened as *Bad to the Bone* blared over the pub's poor speaker system. Patrick made his way back down the aisle mouthing the words to the song. As he bent to squeeze into the booth, his untucked shirt pulled up and away from his pants, Bennett noticed the glimmer of chrome coming from above the waist of his pants. He quickly concluded he had seen the handle of a gun. He decided to confront him.

"You carrying a gun?"

Patrick stopped his mouthing of the words to the song. "Maybe."

"Why?"

"It's dangerous in my world."

"Patrick, you're a bartender in your own bar in New Bedford for crissakes. You're not a pimp on the streets in the Bronx."

"You never know, Bennett. Could find myself between a rock and a hard place someday."

"So you are in trouble."

"Please note the qualifier 'someday'." Patrick returned to his singing.

The singing grated on Bennett's nerves. "Look, I've got to get back to work. Anything else?"

Patrick shook his head, still singing. Bennett got out from the booth. Feeling a tinge of concern, he leaned over the table toward Patrick. "Call me before you even think of using that thing."

Bennett turned to leave the pub. Just before he reached the door he heard Patrick say something, the gist of which he vaguely made out. Something like, "Now that would spoil all the fun, wouldn't it."

The bells on the door jingled as DeWald left the last antique/pawn shop on his list. No harpoons found, and none recently sold. He was nearing the conclusion the perpetrator was more than likely crafting them himself or importing them from outside the area. Actually, for all he knew, the weapon of choice did not have to be a genuine harpoon; it could conceivably be nothing more than a barbed piece of iron rod with an eye at the other end to attach a rope. He scratched the name of the shop off his list and flipped the page. On that page he had written *Tattoo Shops* and underlined it, below, only one name appeared—*The Razor*.

He jumped in his car, started it, and grabbed his cell phone and dialed New Bedford PD. As his phone rang he shifted into drive and started heading for *The Razor*.

"Detective Laykin." The raspy voice blasted through the phone.

"Laykin, it's DeWald. Just checking in." DeWald hung a right.

"How are you doing on this cheery overcast day?" Laykin sounded a little too giddy. Something was up.

"What's happened?"

"We've got us a whale killer."

"Oh?" DeWald knew the list Laykin and Thomas were working on was shrinking. But he found himself surprised they had narrowed it down already. Perhaps he hadn't given them enough credit.

"Thomas paid Mr. Patrick Arnold a visit yesterday."

"And…"

"He set Arnold up and the man fell right smack dab into it. I mean hook, line, and sinker."

DeWald turned left and began looking at business addresses. "How did he set him up?"

"He indicated to Arnold that he had spotted some blood on the deck of his boat—which in fact he had. He further indicated that a lab test might prove the blood to be something more than just your run of the mill fish blood."

"So let me guess," DeWald said, pulling up to the curb in front of a shop with a matching address, "he observed Arnold from afar and the man scrubbed his boat shiny clean."

"You've got it! Well, at least he watched him board his boat with cleaning materials."

DeWald removed the phone from his ear and sighed heavily. His lips moved in silence as he counted. The idiot! He gathered himself and brought the phone back to his ear. "That's great, we know he probably cleaned his boat because he wanted to remove something that could be considered criminal, and hence we can conclude he is guilty of something. Just one thing..."

"Yeah."

DeWald could visualize Laykin shrugging his shoulders. "We no longer have the evidence!"

"So...we know he did it—why else would he scrub the boat?"

"Maybe so, but now we can't prove it!"

"Well how would you have us handle it? You can't have your cake and eat it too!"

"Standard practice, Laykin, standard practice. Get the search warrant, comb his boat, collect the samples, and prove the blood is consistent with that of a whale's or human's with some degree of high probability. If it's human, you do DNA tests."

"And if the blood came back as regular fish blood, where would we be? I'll tell you where we would be, no further along then we are today. A least now we have circumstantial evidence. A bird in hand—"

"Maybe it might have proved him innocent and we could have scratched his name from the list."

"Ah, yes, but now you see, we know."

"Do we really? Maybe his concern was not the blood at all. Maybe his concern was drug residuals might be discovered."

"Ah, come on, you know better than that DeWald."

"No I don't. And even if I did, we have no case against him."

"Maybe not, but now we can concentrate our resources to get the evidence."

"Can we? It could be he wants to hide something else. Besides, he's spooked now. We're going to have a difficult time getting any evidence."

"They always slip up, DeWald, you know that as well as I do."

"For everyone's sake I hope so. So tell me, does this mean Bennett is no longer the prime suspect?"

Laykin laughed. "Note that at the outset of this call I mentioned we had us a whale killer. We still don't have whoever is killing the girls. They're cousins—don't you see they're in it together?"

"How can you say that—if Arnold is killing the whales, how do you know he's not killing the women also? I don't see how you can so easily draw these lines. Don't jump to conclusions, Laykin. I've heard a lot of talking today with little or no evidence to back it up. This is not the time to get our panties in a wad and blow this case because we think we have all the answers. We're a long ways from making an arrest."

"We're damn close. I've let the Mayor know."

"I'm warning you, no arrests, it's too early. Got that?...Did you hear me?...Hello?...Hello?"

When DeWald finally looked at the storefront he realized the address information for the tattoo shop was not really necessary. Tattoo was spelled out in large pink neon letters aligned vertically in the front window. He peered through the glass door and could see a row of what looked like barber chairs. The store appeared to be empty. He stepped inside—no bells on the door to ring in his presence. The right wall of the store was covered with photographs of men and women proudly baring their parts to display the colorful hues planted into their skin. The left wall was wallpapered with colorful designs the customer could presumably choose from. DeWald felt like he was in a T-shirt shop where the T-shirt material was ones' own skin. He scoured the design wall carefully looking for any whale artwork. He found the bulk were of women, some fantasy, something called black and grey, a few tigers, a shark, but no whale.

"Can I help you?"

DeWald turned sharply to face the door where the voice came from. There stood a stout man, his long gray hair tied into a ponytail—his large arms, looking as though dipped in ink—squeezing through the holes in his shirt where the shirtsleeves had been crudely cutaway.

"Can I help you?" the man repeated.

DeWald came to his senses. "I'm looking for the owner of the place."

"That'd be me." He looked quizzically at DeWald. "Just what do you need?"

DeWald pulled his leather badge case from his belt and flipped up the cover so the man could see his badge. "Special Agent DeWald, FBI, have a few questions for you if you have moment."

The big man shrugged and walked over to one of the chairs and plunged his huge body in it. "I was just about to take my afternoon nap."

"Won't take but a moment."

"What kind of information could the FBI possibly want from the owner of a tattoo shop?"

"You'd be surprised. You do tattoos of whales?"

"Whales, huh...don't recall doing one. They're kind of out of vogue, even for the tattoo industry."

"None, ever?"

The man played with his beard. "Well I guess, come to think of it, I gave this one guy a whale tattoo."

"It was guy, not a woman?"

"Definitely a guy."

"What was his name?"

He shook his head and played with his beard. "I can't tell you that. That's client-artist privilege."

DeWald smirked. "I don't think that applies..."

"Well I'm not gonna tell you that—I'm a lot of things, but I'm not a snitch."

"I guess I'll be serving you a subpoena someday."

"You can try."

DeWald didn't like the turn the conversation was taking. He only had a few more questions, but he couldn't afford to have the man shut him out. He decided to end the debate and ask another question.

"You work by yourself?"

"Most of the time."

"Most of the time? What, you have part-timers?"

"I have apprentices."

"Any now or any lately?"

"Mostly teens, so you know what that means, they come and go as they please. Working here is no different than McDonald's. They begin here all hyped-up ready to go, thinking tattoos are so cool, and then a few months later, it's just a job so they quit."

"Any of them stick around long enough to learn the trade?"

"Very few."

"Can you remember any of them?—and these are not clients I might add."

"Oh, I guess I can remember about three or so."

"By name?"

"Yeah, I guess if I thought about it for a while."

"Would you mind writing their names down."

DeWald pulled his notepad from his shirt pocket and handed it and his pen to the man. The man looked up at the ceiling a few times, before finally scribbling something on the notepad. "Best of my memory, probably misspelled some and I could have missed a couple."

DeWald read the three names scribbled on the paper. One sounded familiar. "Any of these guys more talented then the others?"

"One at the bottom. He was good. Still in town I think, don't know what he's doing, haven't run into him for awhile."

Same name DeWald recognized. Interesting.

He walked to the corner of the attic, stripes of moon from the roof scanning across his body. Between his outstretched hands, a few feet of fishing line held taut enough to twang out middle C. He inched toward the corner of the attic, towards the chair, toward—. He stopped and immediately began to shake, and a split-second later his tremors erupted into a violent action of ripping the fishing line from his hands and throwing it to the floor. He had no victim in the barren chair in the corner—no wide-eyed young woman to cast his Ahab eye across. The playing of the role without a real victim had grown as tiresome and dull as masturbation; the time had come for a warm body. He sat in the chair and began to weep. His tears streaked over the makeup caked on his face, reaching his scar and then channeling to his chin. His weeping continued for some time and then he got up from the chair and walked across the attic to his makeup table and laptop.

He spent a great deal of time applying the makeup, morphing himself to Ahab. But moments later he rubbed it from his face into a towel; great globs of flesh tone grease joining dried, crusted semen in the folds of the terry towel. He did not get it all but he didn't care, he just wanted the charade to stop, now that the empty chair had devastated his mood.

He sighed and turned on the laptop. He pulled open a side drawer and carefully lifted a greased stained sheet of paper that had the appearance of having remained folded for some time, and pulled it open from the corner. He squinted as he attempted to make out the photocopy of a handwritten ledger. In the far left column were names, and to the right, columns for quantity, payment, money owed, and product. Some names had check marks next to them. He returned the paper to the drawer.

He returned to the laptop and went to a file listing and traversed it with the mouse pointer until he reached a folder named *Cynthia Mendosa*. He caressed the folder with the hand icon that appeared when he held the mouse over the folder. He nudged it, as if checking to make sure the folder had no life. A smile crossed his face. He moved on to a folder named *Sheryl Bidwell* and caressed it as well. After relishing a memory he moved on to the next folder. He did not caress the folder, but instead spit on the floor and yelled out, "Whore." The folder bore the name *Kathy Rueger*. He moved on to the next folder. He had named it *Felicia Sheldon*. He caressed this one as well and then double-clicked on it to open it and review its contents—a document file. He double-clicked on the document and it opened in his word processor. He rubbed his hands together and began to read:

Victim: Felicia Sheldon
How met: Chat room. She was into actors.
Drug user: Confirmed, in the ledger

Captured:	On 5/24/01. Grabbed her as she walked the beach alone. How stupid!
Whale:	Killed 5/27/01. Humpback, 47 feet, 9 inches. Left at South Marsh Beach.
Ambergris:	Failed. A bombastic couple, wading out in the water in the early hours came upon my whale before I could place the ambergris in it.

Story: She sat in the corner for days. I fed her some and allowed for escorted potty breaks. I was waiting for my whale. My head harpooner, Stubb, is growing careless and undependable. I shall have to deal with him soon, I fear. In the early hours, when word arrived that I had my whale, I went to the attic to have my pleasure. I took great time and spent much effort in decorating my face for the occasion. But when it was done, oh my, how magnificent, even beautiful! Over my shoulder as I worked, I watched in great pleasure as the terror in her eyes grew, grew with each stroke of makeup placed on my face. And when I created the scar, that great scar that spans my face, that great scar that defined the anguish of Ahab like no other body aliment, save for his dismembered leg—she screamed, a muffled scream, but a scream all the same. I was euphoric! When I finished, and arose with great dignity from my chair, and started towards her, I heard trickling water. When I arrived near her, I saw the puddle—she had urinated on the floor. Pip rapped my chest, my innocent limb beseeching me to yield to him, and bestow mercies upon our prey. But Pip's mass, such a small portion of me, and the weight of his pleas, likewise, I was able to chastise him to silence. Returning to my prey, I anointed her, as I had all the others, but her I did more so. I wrapped the cool string about her neck three times and then stood back away from her, still grasping each end of the string. I told her to prepare to die! I jerked the string! And held it and held it and held it. She slumped. I wept profusely, unable to control my joy. But I could not make her ambergris. That damnable couple! I dumped her in the ocean— what could I do? Stubb could not bring me another whale before she started to stink! I have only achieved half of my satisfaction with this woman.

He smiled through most of the text until he came to the end; there his brow furrowed and chin buckled. Anger filled his eyes. He closed the document and clicked on an image file. Up popped a young lady, dead, a whale tattooed on her forehead. He went to the operating system's command line.

```
$ internalize sheldon.gif tchest.c
$ status: sheldon.gif successfully internalized
$ mv tchest.c whale.c
```

The image was successfully inserted into the source file and the source file was renamed to a whale file. He compiled the whale file and then from the command line tested it.

$ harpoon whale

He listed the files in the directory:

$ whale ambergris

Success. Now deliver it.

$ deliver whale queequeg.pequod.net gate.goldengate.net

The whale file was on its way. Destined to slip through the firewalls of both networks, in the guise of an email, snatched away microseconds after reaching *root's* mailbox, by a tiny process wearing the cloak of a common Unix process—the process, having been delivered earlier to the Unix administrator as a security patch to the operating system, complete with supporting credentials: an MD5 hash digest to verify it had not been tampered with, and licensing agreement, to make it appear as though the vendor had sent it. The administrators dutifully installed the upgrade to the operating system, and if they had chosen not to run the process previously, it removed the blocks and brazenly set itself to run again. It continued to perform just as it had before—to report to a user when new email had arrived—with a twist, when it found something it liked in *root's* mailbox, not only did it not report it, it stole it, and then erased any history of the transaction from system logs.

His comrades will be impressed, others, simply baffled. "Thanks, Bennett," he said aloud, heavy on the Ahab accent. The Sheldon folder now contained a whale file, making it complete along with the first three. He glanced at his watch and smiled. Mocha Dick would not check for food for another hour. He could prepare Ms. Sheldon with plenty of time to spare before Mocha Dick checked. By morning, she would join the others in his bowels.

$ echo "update Sheldon.gif clobber" > squid

The command copied the text in between quotes into the file he named squid. Mocha Dick's diet consisted of squid files. When he checked the site that night, he would find the squid file, consume it, and the command it contained would be executed. He would upload the image of the Sheldon girl into himself, replacing the image he had of a young vibrant woman, her sandy blonde hair blown by the wind across her smiling face, with the image of a clay-skinned face, no brain activity behind it, and a whale tattoo etched in the forehead.

He sat content for a moment, but then pursed his lips and pulled open a desk drawer. He reached in and pulled out a photograph of a young woman. He caressed his chin and smiled. Mocha Dick might be hungry tonight; he had not found a squid file for some time now. Perhaps he deserved two courses today. The photograph was gently placed on the glass of a scanner, resting inches from the laptop. The photo was scanned and saved as a GIF image on the laptop's hard drive. He closed the Sheldon folder and went to the command line to type in the command to create a new folder—one to store his new image in. Like Kathy Rueger, he would initiate the ambergris for means other than the purification of drug soiled feminine minds. There were other ways of soiling the mind. And they required a crucible as well. He laughed aloud as he typed the name.

```
$ newfldr
$ Enter new folder name: Hoosier
$ Status: New folder successfully created.
$ echo "insert hoosier.gif" >> squid
```

Mocha Dick was in for a feast tonight.

Chapter XI

The umbrella flapped in the wind above their heads. They sat cooled in the shadow of the restaurant on the peer, the restaurant blocking the sun, sinking on the horizon in the east. Bennett studied Hoosier. Dark sunglasses perched on her soft cheeks—her black hair dancing with the wind. She looked different from her youth in some ways, mature ways, but in other ways she was preserved, her grin for example, she retained. As a child, he fought his parents' edict that he attend the Powerhouse camp and take the daily ride with his mother from New Bedford to get there—that was until the first time he had seen her. She was a young child with a pixy haircut, a pea green bathing suit with a yellow flower on it, rubber thongs, and a puffy Barbie doll beach towel draped over her tiny shoulders. She looked so innocent and scared. Her parents had purchased a summer home on the island and they drove up from Indiana to spend that first summer there. The other children made fun of her of course. She was a bit on the pudgy side, so they heckled her about her weight. But mostly they heckled her about her name. What kind of a name was Hoosier? Nothing derogatory really rhymed with it, so the best the kids could come up with was to replace Hoosier with "who's your." As in who's your daddy, or who's your friend.

Hoosier never seemed to mind. She dealt with it gracefully. One day, when her mother couldn't make it, and her dad came to pick her up, Greg Buxton, a big kid for his age, brazenly asked her dad why he named Hoosier, Hoosier.

"What!" The man wearing a T-shirt and crew cut yelled out. The children were all taken aback. Some ran for the safety of the large rocks at the perimeter of the Powerhouse. "You telling me you don't know what a Hoosier is?"

Greg Buxton trembled out, "Nnnn-ohh"

205

"You've never heard of Bobby Knight and the great state of Indiana!"

The children, wide-eyed, shook their heads quietly in unison.

"The Hoosiers are a college basketball team—the best in this da...arn country." Bobby Douglas, a wafer of a child who wore a bathing cap when swimming, giggled at the man for almost saying a cuss word. "So you know what that means?"

The children again meekly shook their heads.

"That means anybody named Hoosier is somebody very special. It's an honor to bear that name. You all should be so lucky!"

The children never kidded Hoosier about her name again, though not out of their newly acquired understanding as was impressed by her father, but out of fear of her Marine looking father. Later, when in their teens, Hoosier explained to Bennett that while in the womb, her dad was sure she was a boy. Against his wife's better wishes, he demanded that the boy would be named Hoosier, justifying the name on the basis that one day the boy would represent the great state of Indiana at the university. When a girl popped out, it didn't faze him—they had a girl's team too. She never played an hour of basketball, despite her father's constant encouragement and the basketball wallpaper he put up in her bedroom and the goal and half-court out behind the house.

Bennett took a liking to her that first day. It was unusual for him because at the time, girls were not something he wanted to be associated with. But Hoosier had a tomboy edge to her, and Bennett latched on to that part of her. They played together a great deal. Mostly doing boy-like things, fishing and exploring the island. But every once in a while, she could talk Bennett into playing with dolls. Bennett was willing under the condition he could play with his GI-Joe and blowup Ken with a rack of Blackcats. They were inseparable during the summer for a period of years, until the time came when Hoosier began to mature much more rapidly than he. One summer Hoosier showed up with lipstick and eye shadow, Bennett greeting her with a plastic sidearm slung from his belt. Both noted the discrepancy. She was infatuated with boys; Bennett, still intrigued by fetching crabs from seaweed-bearded rocks. Hoosier wanted time on the phone, Bennett time playing games.

Years later his maturity caught up with her. He greeted her that summer wearing cologne and designer jeans. Remarkably the girl who had gone crazy now made a lot of sense. Rather then playing, they worked as counselors at the Powerhouse, talking constantly, sharing intimate details of their teenage lives. One summer, their sophomore year in high school, Hoosier arrived on the island in tears. She had left behind Jeff, an infatuation from Indiana. Three weeks into the summer she had forgotten about him and Bennett forgot about his head injury in the fall. Bennett had kissed her. At first they both looked delighted, but then both of them awkwardly started to talk on some inane subject to distance themselves from the act. They managed to get by the awkwardness as the summer progressed and kissing became a ubiquitous act for their time spent together.

Patrick moved into the picture that summer too. He and Hoosier had known each other, but Patrick had not paid her much attention. As Bennett remembered it, when she arrived that summer, her breasts had ballooned, and when Patrick spotted her in a bikini he took notice. Patrick had said to Bennett he wanted to be part of the gang for friendly reasons, but Bennett thought otherwise. Hoosier told Bennett that Patrick had more to him then Bennett was willing to give him credit. She and Patrick had had long talks and never once did he make a move.

And then, after their junior year, she was gone. Hoosier never showed up at the island the next year. Her parents said she had a good job that paid well in Indiana and she decided not to make the trip that summer. Bennett missed her, but by the next spring, she became a pleasant memory, one of those things filed for life, to be drawn upon in the latter years when one longs to be young again.

But of all the memories they shared, he had one in particular that he drew upon often. It was the summer following his father's death in May. His life had become a singular experience, a monotonous emotion that began every day by waking to stinging pain that by noon transitioned to a vacant numbness that consumed all pleasures haphazardly stumbling into his existence. And then on June 17th he saw her grin. Somehow—a miraculous act he often thought—her grin melted a hole in the stone cold cloud he lived in and brushed living colors on the canvas of his singularly blue outlook. The experience was something he could never forget, and though she left years later, she would remain a living voice in how he perceived his world.

But here she was, sitting across from him, real as life. It felt right, like she belonged there with him. She hadn't touched her food, a shrimp salad. Bennett hadn't touched his either.

Food just didn't seem that important—most foods that was. Bennett suggested they go to the Creamery on the Neck and get a double-dip of strawberry, something Hoosier heartedly endorsed.

When the two were in Bennett's car he caught a glance of his friends in the rearview. "Are they ever gonna cut me some slack, I can't think straight when they're around. I never know when they're going to come in and grab me."

Hoosier kept her look straight ahead. "I know what you mean. You need some peace and quiet to think."

Bennett started the car, heading down Melville, and eventually on Sconticut Neck road, where the Creamery was located, near enough to the top of the Neck not to bother him. He gently grabbed Hoosier's hand. "Can you help me out this weekend?"

Hoosier squeezed back. "What are you thinking?"

"Can you roll your hair up in a ball?"

"Yeah…"

"It's supposed to be foggy tonight. Right?"

"I think I heard that on the radio."

"I have a plan, and it could be dangerous."

She winked. "I'm game."

Hoosier pulled her hair back and grabbed it into a bundle in the back. She looked in the hallway mirror in Bennett's house and turned from side to side. "I don't look too bad with really short hair."

Bennett wrapped the pink rubber band he held a couple of times around his bunched up fingers. He pulled the rubber band from his fingers and secured it around the hair Hoosier had pulled up. "You look great either way." He then went to the front door.

"Why are you going out?" Hoosier asked with a concerned look.

"I want them to see me wearing this hat." He pointed to the ball cap he was wearing. Bennett returned a few moments later. "I made sure they got an eye-full of me. Now, let's have you put on my clothes."

Hoosier smiled, knowingly. Bennett rolled his eyes and said, "You can use my room to dress in."

Hoosier giggled. "If you insist."

Bennett went to his bedroom and returned in shorts, holding the cloths he had been wearing. He handed them over to Hoosier.

She walked to his bedroom and announced she'd need some tape as she stepped inside the door.

"Tape?" Bennett asked.

"Yes, tape. Bring me some masking tape." She said from behind the bedroom door.

Bennett returned with a roll he retrieved from the kitchen. He slid his arm through the door and told Hoosier to grab it.

"I'll need your assistance. And this is no time to be prudish," Hoosier said, ignoring the tape. "Come on in."

Bennett gently pressed on the door and found Hoosier inside with her back to him. She wore no clothing from the waist up.

"I need to flatten myself," she said.

Bennett swallowed hard. "You don't need to take it this far."

"We won't fool them otherwise."

Hoosier had her breasts pressed flat against her chest with her hands.

"Now start the tape and wrap it around just above my hands. As I slide my hands down you move the tape down. A few wraps and we should be done."

Bennett obeyed, aware of his dresser mirror on the other side of Hoosier.

"Good?" Bennett asked, after three or four wraps, his voice cracking.

"Should be good," she said. "You're funny." She pulled Bennett's shirt down over her head.

Bennett felt some disappointment, but was relieved to see the incredible temptation removed. Moments later, Hoosier was dressed in Bennett's clothes. He went to his bedroom and dressed in the darkest pants and shirt he could find. He packed a bag with enough clothes for the weekend. He joined Hoosier in the hallway.

"You look like a short me," he kidded.

"Be nice, now," Hoosier said, waving her finger at Bennett. The cloths appeared baggy, but acceptable.

The two looked at each other, each pondering the magnitude of what it was they were about to participate in.

"Here's your chance to back out," Bennett said calmly.

Hoosier shook her head. "Remember as kids, the exploration on the back of the island? We're in this together."

Bennett smiled. He then, for reasons he never contemplated, kissed her. She kissed him back. The heat began to rise. Five minutes passed before they left their shared embrace.

"Whew!" Bennett said, swallowing hard.

Hoosier exercised her eyes. "I think I saw stars. Not quite the same kiss you gave me that summer when we were sophomores."

Bennett laughed. "You remember that moment too, huh?"

Hoosier wiped perspiration from her forehead. "Almost everyday, for the last ten plus years."

Minutes later, Hoosier stepped out the front door. She turned her back to the street and left a crack in the door. From inside, Bennett put his mouth up against the crack. He talked almost as loud as he could and said he was going to the grocery to get a few things and that he would be right back. Hoosier stepped away from the door and walked to the car. She was glad to see the fog had cooperated. She tried her best to walk like a man; as best she could remember a man walked. Bennett had given her his keys, so she unlocked his car, got in, and backed into the street.

Bennett squinted through a gap in the blinds of his front window as his car backed into the street. He watched as Hoosier drove off down the road. The car out front started up, but the headlights were not turned on. The car began to roll down the street. Bennett breathed a sigh of relief. He was just about to turn around when he saw brake lights. The passenger door opened and a man stepped out onto the road. The car took off, leaving the man standing there. Bennett watched as he quickly darted off into a group of evergreens across the street in the neighbor's yard.

He hoped Hoosier would be able to pull it off. She was to go to the grocery, find a dark spot in the parking lot, one dark enough to hide her features, but not dark enough to hide the movement of her arms and legs. Once there she was to get out of the car and begin walking to the grocery—stop after a few feet, check her pockets and throw her arms up in the air, making it look as much as possible like she forgot her wallet. She was then to step back into the car, drive back to Bennett's house and spend the night and the next day there or even longer if she could.

Please dog—don't bark, Bennett heard himself thinking, as he gingerly made his way through his back yard to the fence. He figured the man across the street was expecting Hoosier to be in the house and would not really expect her to sneak out through the back. He tossed his bag over the fence and winced, hoping the neighbor's dog was safely inside tonight. He heard no barking so he climbed the fence. He ran along the side of the house until he reached the street. There he began a speed walk to Williams Street where he was sure he could get a cab. He wasn't disappointed.

The gravel crunched under Patrick's car tires as he maneuvered his way down a foggy Balsam Street. Caught in the beam of his headlights, a few yards of gravel road and the plush flora at roadside blanketed in a powdery coat of dust from the road. He could not see much further. The meeting was not worth the risk. But his friend, Ahab, had insisted. When Patrick rebelled, he threatened to make it very easy for the cops to pin the whale killings on him. He reminded Patrick of what he had said earlier, that the keyboard was a mighty weapon, and that with a few keystrokes here and there, he could ruin him for life. Patrick sensed that he should believe him.

The gravel road bent sharply to the left. He took the turn with a tight grip on the wheel and, by inches, avoided a bicycler hugging the inner corner, no lights on the bike. Idiot. It spooked Patrick though, something shooting out of the foggy darkness like that. He slowed the car to a crawl. Through his open window he heard the surf of the town beach. He turned on the cab light to verify his gun was resting on the floor in front of the passenger seat and quickly turned the cab light off. The tires hit pavement and he breathed a sigh of relief. He knew only the last stretch to the beach was paved, and he had to be close. A tall chain link fence met the road at either side; the gate was swung open, and kept there by a small sign grounded in a paint can filled with cement. He drove through the gate back onto the gravel parking lot of the beach.

His headlights caught the tower, a large monolithic rectangular block three stories high looming into the night air like a supplanted sarsen stone from Stonehenge. The first two floors were barren of windows, but the third had a sliver of an opening that faced Buzzards Bay. It was a long ago window for the armed forces to make use of binoculars and comb the seas for the enemy while remaining protected. Graffiti covered the first ten or so feet of the tower, sometimes higher if the artist brought something more than a five-foot stepladder. A fairly new piece of plywood covered the only entrance to the tower, the entrance a small square hole at the base.

Patrick parked the car and slinked to the tower. He knew well enough from his youth that the plywood cover was easily pulled from one side and a human could carefully slip through. It was always new, every week it needed replacing because someone busted through it. He hesitated at the small door. He knew the illegal drug folks wouldn't be inside; it was a trap, because if the cops showed up, there was nowhere to run. Drinkers or lovers could be inside. Lovers he didn't worry about, there might be a moment of embarrassment, but a brawl was not likely to break out. Drinkers he did. He could not forget about his friend Ahab either, it might be a setup.

He rapped on the wood and heard his knock echo on the other side. He patted his side, felt the solid mass of his gun. He pulled the gun from his belt and pried back the plywood. He slid his weaponless hand in first, knowing better than to use the other hand where the weapon could be easily wrestled from him and used on him before he knew what was happening. Damp, stagnant air rushed over his arm, and he had the sensation of a residue settling on his skin. He popped his head through. The stench—a potpourri of urine, stale beer, must, liquors, vomit—almost overwhelmed him. He pulled his knees to the edge and quickly pulled back his hand he had placed on the floor for balance. Moments later he stood in the center of the tower looking up towards the top where a thin sliver of moonlight banded the inside wall. He could make out a rickety iron rod spiral staircase winding to the top floor. He took his first two steps and shook the railing checking for stability. Classy joint. He heard loud creaking and the railing move very loosely in his hand. As he looked up at the top floor, a grid-like layer of metal, he noted a black spot in one corner. Someone was up there.

"Hello?"

Silence. He adroitly made his way up the stairs, keeping his eye on the spot. He watched it move. His assumption was confirmed. He continued with his creep like pace until his eyes were at floor level. There he saw the figure, leaning against the wall, peering through the only window the tower had.

"Ahab?"

The figure turned from the window. "Stubb, you're late."

"That's Patrick."

"Quiet Stubb, I know who you are."

"It was the fog, had to drive slowly."

The saggy, traumatized face of Ahab, looked more distraught then ever. "I'm not happy with your skills, Stubb."

"Look, I don't know where you are going with this but let me tell you now, I'm out, and I mean out." Patrick watched the jagged scar on Ahab's face flex, then quiver.

"What's this? Mutiny!" Ahab hissed.

Patrick shook his head rapidly, not believing what he was hearing. "Mutiny? No, its not mutiny, it's not safe—the cops are crawling all over me. Some freakin' battle cruiser with Greenpeace crazies is on its way too! No amount of money is worth it."

Ahab moved from his corner position, the moonlight banding his eyes, illuminating the surrounding gray harried skin, as he peered out to sea. "The wretched must be purified. Imbibing in chemical substances to alleviate or alter the pains of this cruel world, soils their soul, and without proper cleansing leads to an unholy stench, a fetid aroma that in time consumes one to such a degree that it can acclaim infinite persistence, not only to the one plagued, but to those connected by the umbilical of fate. We must purge this world of such reprehensible behavior by providing violators a new beginning, a reclamation of the womb, and symbiosis with a surrogate organic cavity that they might reappear cleansed in the sweet scent of purity." Ahab stood in silence and spoke without turning his head. "Ye hear me, Stubb?"

Patrick stumbled for something to say. "I have no idea what you're saying. What does it have to do with anything?"

"The mighty whale is the only crucible capable of bringing about the cathartic process I have been explaining to you." Ahab turned from his view of the sea and looked directly into Patrick's eyes; the band of moonlight exaggerating his scarred face by producing jagged shadows on his skin. In a stern monotone voice he said, "I need a whale, Stubb. What say ye?"

"They're watching my boat."

"We'll take mine. I have one."

"It's not worth it, I don't want to go to prison."

"For hunting the leviathan? Please, it's a God given right. I have already informed you, in some great detail I might add, what the consequences of your disobedience will buy you. Attend this last hunt with me and there will be no more."

"Why should I believe you?"

"Does it matter?"

Patrick sighed. "I won't do the harpooning."

For the first time the cracked corners of Ahab's mouth turned up and something that approached a smile awkwardly embellished his face. "That pleasure's all mine."

A thin puddle collected in the seat of a blue plastic chair, placid, like a round piece of glass. In fact, all the chairs had a puddle, including those lining both sides of the ferry, those clustered up in the bow, and the ones evenly placed in rows at the stern. The dew point rose through the night, and moisture settled and collected on everything. Bennett included. He stood below deck, looking out a porthole at the deck of the Island Ferry—shivering, his clothes permeated with dampness. The spur of the moment planning he conceived the night before failed to take into consideration lodging for the night. He realized this when he entered the cab. He knew he could have easily gone to a hotel, but the night had acceptable sleeping conditions for outdoors, and he knew of benches near the dock for the ferry and he figured a night on one of them would not be too bad. He had miscalculated.

He was the first one in line that morning. Drawn more by the thought of a hot cup of coffee he might purchase at the ship's tiny café, than by the thought of getting on the ferry as fast as possible. He kept an eye out for those assigned to follow him, looking over his shoulder, all the while wondering what he would do should he spot them. And now, on board, one cup of coffee already inside him, and the thought of another, close to his heart, he would not let down his guard, watching through the carefully selected porthole with a keen eye. Each head appearing first at deck level, as individuals boarded the ferry by ascending a plank, caused a hesitation in his breathing. Families, the elderly, lovers, bicyclists wearing spandex—a representative from every socio-economic group, walked up the plank, but none bearing any resemblance to Bennett's followers. At last, the gate on the ferry closed; there would be no more passengers.

Steam from his next cup of coffee fogged his sunglasses, but he refused to remove them. He slid them out on his nose to allow more air to pass between them and his face and they cleared up. He looked odd wearing them; the skies were gray with pregnant clouds, hovering very near to the ground—but he knew if anybody spotted him…He pondered whether he had done something stupid, something rash, something that might expedite the actions of the New Bedford PD. The DA would be involved soon, if not already. Even with DeWald presumably on his side, he probably could not fight city hall. He knew one thing—he would call DeWald the minute he arrived on Nantucket. Forgiveness was the easier thing to ask for than permission. Perhaps if DeWald knew where he was he could circumvent the knee jerk reaction guys like Laykin and Thomas were sure to have.

The ferry horn blasted three times, and Bennett felt the boat begin to move. He felt comfortable on the ferry. Many times in his youth, and not so many in his latter years, he had taken the ferry to visit his Aunt Grace. But this was the first solo trip and certainly the first surprise visit. Aunt Grace would take it well though; he knew that to be the case, her gentle demeanor dictated it would be that way. He pictured her small two-bedroom cottage on Davis Lane. For as long as he'd been alive, she had lived there, an old maid, rarely venturing from the island. She was his grandmother's sister on his father's side. He recalled the picnics in the backyard, how one time he had wandered off, following Osprey Way down to a large salt marsh pond, his father yelling for him and for some strange reason, him ignoring the calls. He never forgot the worried expression on his father's face when his father found him squatting at the edge of the pond, watching a school of minnows glide along the shore. He cherished the memory.

The ferry pulled out of the bay, heading directly for Woods Hole. Miraculously, the sun burned through the hazy skies, pulling Bennett from below deck to a chair on the deck. The warmth felt good on his skin, and the restless night before snuck up on him, his eyelids gradually clamping shut, and his body succumbing to a deep sleep.

"It's risky meeting here," Stacey said, taking a seat in a cushioned chair that made up part of a set of patio furniture on her back porch. A tall wooden fence enclosed the backyard, making it somewhat private.

DeWald ran his hand up and down his glass of iced tea, rubbing condensation from the glass onto his hand. He took a sip. "It's safe enough. I'm tired of driving all over the map to talk to Tanya."

"Bennett knows I live here."

"Now you tell me."

"You were the one who phoned and said you were coming over, never giving me a chance to reply."

"My mistake. But we're here, so let's make the best of it. What have you learned about Elander?"

Stacey began rocking in her chair. "Elander is an interesting character."

"Interesting? In what kind of way?"

"The way he carries himself, speaks, and so on. He's bizarre, really—to put it mildly. He acts like he's a modern day Confucius. He's into the cyberspace stuff big time. To hear him talk, one would think his citizenship were there instead of here in the states."

"What's your gut feeling?"

Stacey cocked her head. "He could be a fit for our cyberpunk, but he lacks the fortitude to kill I really believe. He wouldn't kill a flea, let alone a whale or woman."

"But could he hire someone else to do it for him?"

DeWald watched Stacey's eyebrows rise over her sunglasses. "Perhaps…if there were some weird way it tied in with his beliefs. But on the surface, it's incompatible." Stacey swatted an unusually boisterous fly from the top of her glass. "Why the interest in Elander?"

"He was picked up yesterday."

"For what?" Stacey asked, slack jawed.

"Selling dope. Apparently selling pot is not against his belief system. He told the detective who booked him that he believes he has a constitutional right to sell it."

"Well I guess selling pot doesn't make you a murderer." Stacey meandered over to a small garden at the edge of the patio. She pulled a weed from the bed and tossed it into the yard.

"No, but carrying a black book that contains the names of two of the three victims might dispel that line of reasoning."

Stacey swung away from the garden to face DeWald.

"The arresting cop found the book in his back pants pocket," DeWald continued, "seems he kept good records on his customers, who they were, what they paid, how much they ordered. The Sheldon girl reported missing last week—she's in there as well."

Stacey returned to her seat. "How many total customers did he have in his book?"

DeWald shrugged. "Does it matter?"

"I'd say so. If he has two or three hundred, then having the names of two victims isn't as significant."

"True. Regardless, it's circumstantial. We need to pursue it and see if there's anything to tie him directly with the victims." DeWald polished off the last of his tea and began working on a piece of ice, pausing for a moment to ask, "What about the kid, Adam?"

Stacey took a sip of her tea and swished the fly from her face. "I haven't spent much time with him, but I'm almost one hundred percent certain he has no involvement. He's just a kid really, into his work at Pequod, loving every minute of it. I can see nothing that would indicate any trace of violent behavior."

"Collin?"

"Him, I don't know about." Stacey sighed. "He's not one I really want to pursue."

"It's your job as Tanya, Stacey."

"My intuition tells me that if I show any interest in him, he'll be one to latch on, like a lost puppy."

"Use the Bennett angle."

Stacey began rocking in her chair. "What do you mean?"

"Tell him you're concerned about Bennett, and he being Bennett's best buddy, he's the most likely candidate to talk to about it."

"I'm not sure I can do that."

"Why?"

"I don't want to perpetuate anything, I want to cut it off. I don't want to hurt Bennett."

DeWald was surprised to hear this—rumor had it Stacey was driven—the type not easily detoured from the trail. He chuckled. "You may not have to concern yourself with that anymore."

"How's that?"

"An old friend of his is back in town. The guys assigned to him say they're getting pretty chummy with one another."

Tanya slowed her rocking. "Oh…"

"It's the best thing that could have happened Stacey, you know that."

"Of course you're right." She shifted in her chair. "So…yeah, it's for the best…so what have you learned?"

"I think there's somebody we need to start focusing in on. He worked as an apprentice with a man running a tattoo shop."

"So who's behind door number three?"

A screaming baby, accompanied by an apologetic mother, snapped Bennett from the dosing he enjoyed in the chair on the deck of the ferry. He jerked in reaction to the scream, sending his sunglasses to his feet. For a moment his sudden arrival to a conscious state was disoriented in the void between his dream and the reality of his environment. He collected himself as best he could and began looking around. From the stern he saw mostly blue seas capped by a wafer thin sliver of land, which had to be Martha's Vineyard. He knew the island must be close, so he collected his bag and the laptop he purloined from work and proceeded to walk in the direction of the bow. As he made his way, the shores of Nantucket slid along by the other side of the boat. Bennett sucked in the briny air, hoping the oxygen might bring clarity to his fogged mind.

Nantucket invoked a special aura for Bennett. The island, steeped in an odd blend of whaling history, garden cottages, and celebrities, carried forth with great dignity, almost royal in nature—the colonial architecture, crowned with widow's walks and carpeted in cobblestone streets, somehow circumventing the collective will of time and commercial prerogative. The yards and gardens approaching the magnificence of those so impeccably cared for across the pond in England. Nantucket, such a sweet place. The ambience invigorated Bennett, causing his mind to clear and his goal for the weekend to move to the forefront of his thinking.

Minutes later the ferry inched its way to the dock. A section of railing was removed from the deck and a plank abutted against the boat where the section was removed. The passengers poured down the plank, onto the dock, and into Main Street. Bennett stopped at the corner of New Whale Street where a bicycle rental shop nestled between an ATM and a T-Shirt shop. His fellow passengers had already formed a line and it was nearly half an hour getting a bike. The trip to Aunt Grace's was a little over three miles. He calculated he should arrive in about the same amount of time it took him to rent the bike.

The ride took longer than Bennett expected, mostly because Main Street was extremely crowded with shoppers and tourists. But once beyond Main, he made good time down Cisco Road. The road up to Aunt Grace's house was dirt, but relatively smooth. Still, by the time Bennett reached for the doorbell on Aunt Grace's porch, he felt himself shaking from the vibration of the ride.

"Bennett? Is that you Bennett? What a surprise." A disembodied voice sprang from behind the nearly opaque screen door.

Bennett replied back with the most enthusiastic voice he could muster, but apparently it wasn't enthusiastic enough.

The door swung wide. "What's wrong?" Aunt Grace threw a feeble hug around him.

"A little tired, long journey an all."

"Well this is such a wonderful surprise. We don't see enough of each other anymore, not since your father died."

Bennett nodded. "I have a lot of work I need to get done. I made a spur of the moment decision to visit you out here where it's quiet." He smiled. "And, of course visit my favorite aunt also."

"Well, where are my manners? Come on inside. Let's get you something cold to drink."

Bennett agreed—he sure could use something. The bike ride took more out of him than he expected. Once inside the kitchen he spotted the golden harvest colored phone mounted on the wall. It had a rotary dial, and his guess was his aunt was still naively renting the thing from the phone company. He asked if he could make a quick call to shore, to which she dropped a hand and said of course. He desperately wanted to call Hoosier, but knew his home line was probably tapped, so he fought off the urge. He pulled DeWald's business card from his billfold. It was Saturday and he was concerned he might not reach him.

"Where did you say you were?" DeWald's distraught voice rang out through the phone.

"Nantucket—look I needed to get away—"

"Damn, Bennett."

Bennett caught some muffled cussing. DeWald must have put his hand over the receiver, not realizing Bennett could still hear him. Bennett waited for the cursing to stop before speaking. "I'm sorry Agent DeWald, but I've got to tell you, my butt—" He looked over at his aunt who turned away and began washing things in the kitchen sink. "—I need to help myself out. I think I've got better odds of getting this thing solved then you do—no offense. But I need some peace and quiet to work the issue. I can't think while I'm wondering when I'm going to get hauled—," thinking over what he was about to say and stopping himself, "...in that environment."

DeWald sighed heavily. "Bennett, I don't disagree with what you're saying, but do you realize what your strategy looks like to law enforcement?"

"Of course."

"You should have talked to me before you ran off."

"It was late—there wasn't time."

"You could have left today."

"And you would have let me?"

DeWald paused. "...At least tell me what you're going to do."

Bennett looked at his aunt and could tell she was listening. How could she not? He asked if he could move to her bedroom phone, telling her the conversation was kind of private. She told him it was no problem. When he got to the bedroom, his aunt hung up the phone in the kitchen, he never doubted she would—she was a godly woman. He carefully explained the plan to DeWald; DeWald appeared to like it.

Bennett wrapped it up saying, "But I'm going to need time."

"I'm not sure how much I can give you. Your actions today will probably set into motion a boatload of knee jerk reactions."

"Look Agent DeWald, I'm the only one who can trap this guy. You lock me up, and nobody gets him. I know you've got good people in the computer crimes unit, but you and I both know I'm the only one who can catch him. You scratch my back, I'll scratch yours."

DeWald knew Bennett was right; the talent pool in the computer crimes unit couldn't compete with the talent available in the commercial world, where money and cutting edge technology kept them there. "How much time to put this plan of yours into action?"

"A week, maybe two."

"I'm not sure I've got the clout. I'll do what I can."

"It's for both you and I, and the victims."

"I'll take care of explaining your hasty departure. It won't be easy coordinating this; I'm back in Boston visiting my wife. I'll tell them I agreed to it up front. But you're going to have to meet me half way."

"How so?"

"I'm going to contact Agent Kelsey, he's on the island for the week on other business. I'm going to ask him to keep an eye on you—giving you some breathing space mind you. He'll stay out of your way. But if I don't have someone watching you, they'll have a warrant for your arrest today, so I think you'll agree to having Agent Kelsey around."

"I'll live with it—guess I've got no choice."

Bennett showered and Aunt Grace prepared a lunch for the two of them, which she served on the back deck. Her rose beds looked as beautiful as ever. The woman could certainly garden. Bennett couldn't recall the last time he had tasted such a good ham and cheese sandwich. He worked the garden fresh vegetable platter down to a single celery stalk. Aunt Grace asked for a second time if he were in any kind of trouble, and again Bennett replied in the negative, suggesting his phone call earlier in the day was a business issue with a coworker. Aunt Grace accepted his answer and informed him she was going to make a visit to the market and asked if Bennett would care to join her. He declined politely, bringing up the inordinate amount of work that was awaiting him.

Once his aunt departed Bennett setup his laptop on a small shaker table facing a picture window with an unabated view of Pear Tree Cove. He sank into the cushy office chair and found himself staring out at the cove. He watched a group of children in brilliantly clad bathing suits frolicking on the shores. How young they looked, how old he felt. He knew the odds were greatly against him catching the fleeting Mocha Dick. Like the universe, cyberspace grew daily, expanding, creeping across all corners of the earth. Mocha Dick could be out there anywhere, and where he was hanging out today, was not necessarily where he would be hanging out tomorrow. But physically, the killer, the face behind Mocha Dick—he was near—he had to be. Physically he could be in the cottage across the street, but in cyberspace his creation could be sucking up CPUs on a server in Russia. Bennett knew he had to find his creation in cyberspace first, and next track him to his physical location—not an easy chore.

But, the plan was to have Mocha Dick come to him. He removed a computer magazine from his bag and flipped to the back pages. It wasn't long before he struck the hundreds of ads placed by service providers wanting to host a web site for him. He winnowed through them, writing down the ones that came across as residing on the left side of the political equation. Ideally, he wanted a hyper-free-speech type of provider. He called and hit pay dirt on his first provider.

"I don't give a damn if you put naked pictures of my mother up— we're in the web hosting business, not the censorship business," a gruff voice replied.

The price was cheap enough too, so Bennett signed up. He copied down his account information to allow him remote access to setup his site from the comfort of his Aunt Grace's cottage. As he dialed into Pequod, he turned his thinking on content for the new site. He had given it thought earlier, but now that the time was upon him, he felt himself second-guessing his earlier ideas. He felt like a fisherman, kneeling down and peering into a disheveled, malodorous tackle box loaded with every kind of shape and color lure imaginable. What would the fish bite on? What looked good to him, a fish might find laughable—a crapshoot, really.

His modem connected to Pequod and he logged into his Unix box to check the logs of *dory*. He checked the search results log first.

Matches found with high probability:	130
Matches found with medium probability:	1806
Matches found with low probability:	6104

Dory had slowed down its pace drastically. Only two more high probability finds. It didn't make sense; up until now *dory* had only navigated a only tiny bay on the sea of the Internet. He checked to see if *dory* was having some sort of trouble. He looked in the system log.

[dory] Caught signal INT
[dory] Exiting

He checked the *uptime* on his machine and found the box had been running for weeks. Somebody had explicitly killed *dory*. He checked the audit logs, finding them clean. Not much of a surprise. As a consequence of this latest attempt at sabotage, he pulled a floppy from his laptop case that contained the source code for *dory* and altered the code so that should somebody try to kill it again, he would be notified by e-mail. He also changed *dory* so that the next time somebody told it to die, it would rename itself. If the operator were to check if *dory* died by searching on its name, dory would not be found; perhaps fooling the perpetrator into believing the kill was successful.

A single piece of mail was delivered to his fraudulent mailbox for Frank Gerald.

Frank,

I don't know what type of work you were considering with the White Whale, but I may want a piece of it. I visit the bar regularly...

A dead-end, but no surprise, there would be a lot of them. He looked at his watch, too early to try to reach Hoosier via net phone. That would happen later in the night. They had picked up her laptop on the way back to the house the night before. Bennett had hastily setup an account for her with Pequod. He knew the cops could listen in on his analog phone lines at the house and hear voice, but bits and bytes going down the line was an entirely different story. And even if a sniffing device captured the communications, they would be encrypted and would stay that way for some time to come. So Bennett had little concern for eavesdropping when talking to Hoosier on a net phone as long as she talked quietly, something they had gone over before his departure.

Once his browser was up, Bennett began searching for information relating to serial killers—and was amazed at what he found: an entire community of sites dedicated to proliferating facts about serial killers, down to last day and time they killed, and what they ate that day for breakfast— Manson apparently having a bowl of Quisp. He found tables, charts, graphs, shoe sizes, pet peeves, eye and hair color, place of birth, height, weight, favorite TV show—*favorite TV show*? The sheer volume of data was ridiculous. In an odd way he found some comfort in it; he would not be the only one with such a site—like setting up a porn shop, knowing there's another one down the street—and there was plenty of content to chose from. He found a web ring, a series of sites that link themselves together in chain-like fashion so that by jumping in at any particular site and following the links one goes full cycle, visiting them all. He observed that joining would be easy; all of the participants provided a form to join the ring.

After an hour or so a home page was completed. Headed by a sparkling marquee, followed below by blinking stars that appeared to be rotating around photographs of the killers. He called his site *Death Row*. He built a long list of links that would take the user to other related sites, including the ring. He brought up graphics software on his machine and drew a blank square with a big red question mark in it. He pasted it on his home page and then placed some text below it.

New Bedford strangler—I want to party with you, man. I want some of you! You're the baddest out there! The other geeks on this page pale in comparison to you. Drop me a note, man, let's party! Send me some photos—I'll put them up, no lie! I'm for you, man. Let's rock!!!!

He found himself appalled by his own writing. But when bait fishing for bottom feeders, the more rancid the bait, the better. He had one last step: register with all the major search engines. He spent forty-five minutes in doing so, logging off from Pequod and shutting down his laptop afterwards. He glanced out the window, the children were gone and the sun was setting. Having the bait in place made him feel better. Now came the most difficult part…waiting for the bobber to agitate.

Bennett bounced wildly once the tires on his bike made contact with the cobblestone streets of downtown Nantucket. Although it was the best food he had been in contact with in days, Aunt Grace's Yankee pot roast weighed heavy on him. The bouncing did not help the situation. He found a bike rack and dismounted, locking the bike to the rack. The ride to downtown Nantucket, despite the indigestion, was relaxing. He took a deep breath of fresh Nantucket air and gradually exhaled. A fog had settled-in creating a halo-effect around the colonnade of black decorative street lamps.

Down the street Bennett spotted a pub and headed in its direction. He stopped at a cobblestone intersection on the way and, catching odd stares from those who passed by, waved up towards the top of a three-story building. He knew Nantucket had a 24 hour web cam mounted there and had told Hoosier to watch it at about this time so that she could see he was all right. He eventually crossed the street and walked towards the bar. When he got there he glanced through the frosted glass doors at the entrance. From what he could see, it looked fairly tame, something appealing to him. He wondered where his shadow might be, DeWald had said there would be an agent assigned to him, but if so, the guy was a chameleon. The wooden floor creaked as Bennett stepped inside. The pub was small but comfortable for the few who chose to collect there. Straight-ahead, a collection of wooden tables and chairs, while at the back, a pool table, two men playing. A long wooden-top bar ran along the right wall. The bartender behind it was leaning on the bar and talking to a man seated on one of the many empty stools on the other side. The walls were lined with whaling objects and pictures—definitely Bennett's kind of bar—replete with pallid aged paintings of men at the oar on high seas with the harpooner poised to thrust the harpoon he held high. In the backgrounds majestic ships in full sail, awaited their crew to return with fresh spoils.

He moved his gaze to the other wall where the harpoons were mounted on elongated oval wooden plaques. Something inside him rendered the objects things to marvel. He ogled them with a high degree of fascination. Moments later he moved his stare to behind the bar, an ivory-colored, conical piece of wood rested between the beer taps and the bottles of liquor.

"That'd be Ahab's prosthesis," the bartender said with a hearty laugh.

"So you got to him before the white whale?" Bennett fired back with a chuckle.

"That we did boy, that we did. Had a hell of a time talking him out of it. What can I get you?"

Bennett took a stool one down from the other man already seated and replied, "Whatever you've got on draft."

The bartender placed a beer-filled glass in front of him. Bennett took a couple of gulps. He looked over at the man seated next to him. The man wore a ball cap with a company's insignia on top. His hands and face, while clean, were worn ragged. Bennett recognized the signs of commercial fishing wear and tear.

"Was this bar in existence back in the whaling days?" Bennett asked the bartender.

"Nope," the bartender started matter-of-factly, "this humble place of mine once was part of a hotel. All this stuff in here is mostly fake—just here for the ambiance."

"Even the harpoons?"

"Oh yeah."

"They look real."

"Nope. I get them from some guy who sells them outdoors up there at Quincy Market in Boston."

"I wonder if anyone still uses these things?"

The bartender looked puzzled. "What for? It's illegal to hunt whales. You'd have to have a screw loose or something."

"You haven't heard the news about the whales they've found harpooned in the Buzzards Bay area?"

"It's all a bunch of baloney."

The other man at the counter with Bennett cleared his throat.

The bartender laughed and said, "Unless of course you believe Fred's crazy story—"

"It ain't crazy!" the man spoke in an irritated, raspy voice.

The bartender shook his head and looked at Bennett. "It's crazy—"

"It ain't crazy!" the man repeated.

"All right then tell the story to?" the bartender looked at Bennett.

"Frank," Bennett replied.

"Tell the story to Frank. If Frank says the story isn't crazy, then free beers for you the rest of the evening."

"All right," the man replied back without hesitation.

Bennett turned in his chair to face him.

The man began, "I'm a fisherman—not for hobby, but a commercial fisherman. I stay out three weeks at a time. We cruise all over. Well, about a week ago we were a good ways up north to net some cod. It was close to dusk when I left the dining galley for the deck and a cigarette. Everyone else stayed below—"

"How convenient," the bartender interrupted.

The man stopped and stared at the bartender.

"Oh go on…" the bartender chided.

The man took a gulp from his beer. "As I was saying, I was the only one on deck at the time. So I lean on the railing sipping my coffee, smoking my cigarette, and looking at a beautiful sunset. All of a sudden I see we're surrounded by a herd of whales. Now, this is of no great concern because this happens all the time. I enjoy their company quite frankly. So anyway…" A gulp from his beer. "So anyway, I'm enjoying watching the whales—there were mothers with their calves, no bulls. All of a sudden I see this whale come alongside the boat, fairly close, you know. But the odd thing is, it looked as though something was sticking out of its back—not too high either, but sticking out just the same. As I looked closer, I see there's a rope attached that appears to be dangling behind—"

"Now here's where it gets crazy," the bartender interjected.

"Please!" the man said curtly. "So, I notice the whale picks up speed and begins passing by me. But as she does, she gets closer to the boat. I could almost reach out and touch her. So, she pulls out ahead and I notice this rope, attached to the thing sticking in her, is trailing behind. So my eyes follow this rope—and I'll swear to the Bible," he placed his hand on his heart, "when I reach the end of the rope, there attached to it, is somebody's arm!"

"You're so full of it Fred!" the bartender burst out.

"…And I saw it so good I actually saw a tattoo on it—I swear on the Bible!"

"What did the tattoo look like?" Bennett asked with great interest.

224

"Looked like a whale," the man said.

The bartender began another burst of laughter. "Maybe...maybe..." he said, attempting to talk in between his laughing. "...Maybe, it was Ahab!"

The man defiantly shot back, "Kiss my—"

"Draw it for me," Bennett asked impatiently and slid a cocktail napkin to the man along with a pen.

The man took the napkin and did his best to draw a whale. He then drew the letters *S-E-H* near the whale's belly.

"What are those?" Bennett asked.

"I don't know, but they were there," the man said.

"What do you say Frank? You believe this old coot?" the bartender asked.

Bennett thought for a moment and laughed. "I'll tell you this much—it's a whale of a tale if I ever heard one, but I suppose it's possible." Bennett took the napkin from the man, folded it, and stuck it in his shirt pocket.

"Bah..." The bartender said, waving him off.

When Bennett arrived back at the cottage, he found his aunt ensconced in her burnt orange, brown, and yellow plaid recliner—the one he recalled so vividly from his youth, the marquee of her home—she was deeply intrigued by a novel. She did not notice him as he walked by. He tried his best not to startle her, but ended up doing so when telling her he was going to work on his laptop for a while. She jumped, but settled down quickly. Bennett apologized and sat down in front of his laptop. He removed the napkin from his pocket.

Although he had decided that his detective skills lie in the cyber realm, it couldn't hurt to toss in physical evidence should he come upon it. He thought about the old man's story and knew there had to be a connection. If true, did it mean Ahab/Mocha Dick/the killer, was dead? Couldn't be—he had received email from Mocha Dick after the man's story supposedly took place. But it could be timing—the old man's could be off or he could be off. If the killer was dead, that could actually be bad news. The killings would stop while Bennett was under surveillance—it would be simple to draw the wrong conclusion. He shook his head and studied the napkin.

He decided the drawing might mean something to someone. He opened a drawing package on his laptop and carefully began to recreate the napkin drawing using the software. After a few attempts he was satisfied with the look and he saved the drawing to a file. He opened up the web page he created earlier in the day and made space for his creation. It looked odd and out of place, but the possibility that it might draw some attention convinced him it needed to be on the page. He added the text:

"Found a radical tattoo. Anybody out there recognize it? I want one just like it, done by the same guy."

"Are you doing okay, Bennett?"

It was Bennett's turn to jump. He was so caught up on creating the web page that he had not noticed Aunt Grace walking up to him. She patted him on the back. He noticed Aunt Grace staring at the napkin resting by his laptop. "You've always had a fascination with whales haven't you?" she asked.

"For whatever reason," Bennett said.

"When you were a boy, you enjoyed visiting the Whaling Museum with your grandpa."

"Parts of my youth are sketchy. I lost a lot of memories with that accident," Bennett said staring at the floor.

"But you do remember the whales…"

"Yes, for whatever reason, I do."

Aunt Grace gave a serious look. "How is your epilepsy? Is it under control now?"

"For the most part. I have an occurrence every now and then."

Aunt Grace shook her head. "My sister had it something awful. Just awful."

"Yeah, Granddad was telling me about that the other day."

"Of course back then it was more of a mystery. There was no way to stop it. I remember when a spell would hit, your granddad and I would stand by and make sure she didn't hurt herself."

"So, my grandmother, did she have it most of her life?"

"Yes, I believe she was born with it. She had blackouts too."

"Blackouts?"

"Yes. I remember times when I would go into town with her and people would speak of having had conversations with her, which she later told me, she had no recollection of. They said she acted differently in an odd sort of way. Coincidentally these were the visits she could not remember."

"Did she drink?"

Aunt Grace looked aghast. "No never. She was a Christian woman."

"So what you're saying is she did things that later she could not recall ever doing them?"

"Precisely."

"I hope that doesn't run in the family." Bennett rubbed his forehead.

Aunt Grace studied Bennett for a moment. "I can recall you rubbing your forehead the first day you sat up in that hospital bed two days after your accident."

Bennett shook his head. "I'm glad somebody does."

"You had us very worried for two days. What an odd injury. I just don't know how you could have gotten it."

Bennett nodded. "I don't know. Sometimes it scares me that I don't know."

Aunt Grace smiled. "You're a good kid Bennett. It really doesn't matter what happened back then, you've made your life a success."

"I know Aunt Grace, but what haunts me is I lost something special that night."

Aunt Grace gave Bennett a surprised look. "What makes you think you lost something?"

"I'm struggling with a puzzle—and I shouldn't be."

"Shouldn't be what?"

"Shouldn't be struggling. I should be able to figure it out."

"Bennett," Aunt Grace said and put her hand on his, "be patient with yourself. It's good to strive to be the best at what you do, but at the same time, every once in a while, it's okay to look back and see how far you've come—you've come a long way, Bennett."

"Sure. It's just there are so many unanswered questions about the accident. Why me? What did I do to deserve becoming mentally crippled?"

"You're not mentally crippled."

"I know…I know…poor choice of words. It's just the issue seems like a bigger deal now. Right now, more than any other time in my life, I need to be very smart."

"I don't know what it is that has brought about this urgency you speak of, but I do know this—you will work your way through it. You always have and you always will." Aunt Grace turned her head and looked over at a photograph of Bennett when he was six years old, standing in her living room aside a towering construction of Tinker-Toys. "I've watched you for all of your life Bennett. You've always been special, and I don't mean just to me, you're a special sort of person."

Bennett blushed and said, "Thanks, Aunt Grace you've always known how to lift my spirits."

Aunt Grace went to bed, leaving Bennett to himself in the living room. He walked over to the desk, turned on the laptop and dialed into Pequod and checked to see if Hoosier was online. He breathed a sigh of relief when spotting her active connection. He wrote down her address and started up his net phone. When the device appeared graphically on his screen, he typed in Hoosier's address and pressed the send button. A ringing sound came from his laptop's speakers. After five rings, he heard a faint, broken up voice on the other end.

He leaned into the laptop to get closer to the built-in microphone. "Hoosier, is that you?"

"Bennett?"

"Yes, it's me. Are you okay?"

"Yeah, just fine. Haven't been bothered by anybody. I think it worked."

"That's great."

"Saw you on the web cam."

The message: *network congestion*, displayed on the net phone's status bar.

"How…going…aunt's…" Her voice was breaking up.

"Repeat…"

"How are things going at your aunt's?"

"Pretty good, I've got things set up."

"That's great!"

"You can leave tomorrow if you like. The cops know I'm here. They know I left yesterday."

"They know?"

"Yeah, I called DeWald and told him."

Network congestion.

"Okay, I…tonight…"

"Repeat…"

Network congestion.

"I hear…somebody…house."

Bennett's heart began beating rapidly. He couldn't make out much, but he thought he detected fright in her voice. "Repeat…"

"Somebody's in the house!"

"Hooiser, get out of the house, you read me?"

Empty static.

"Hoosier, you read me?"

Empty static.

"Hoosier!"

Chapter XII

Thomas parked himself in one of the orange plastic chairs scattered about the checkered linoleum floor of the break room. He popped open the cold can of Mountain Dew clutched in his hand and took a gulp. He rested the can on the round table in front of him and propped both legs on another chair across the table. Leaning back he looked out the large glass window in the break room, out into the next room where Laykin and DeWald stood toe to toe. He quietly observed the clash, not able to hear a word through the thick glass. Laykin, consciously gesticulating madly in sync with his shouting, unconsciously, the veins in his neck protruding, his face bright red and DeWald, slightly calmer, poised, countering Laykin's wild gestures by decisively pointing down at the floor to emphasize his speech. Thomas sighed, looking around the break room, wondering why they had to have both a Coke and Pepsi machine. He pondered the mindless thought for a moment, gave up, and returned to watch the action next door.

Round two started. DeWald now stood at the white board, listing something Thomas couldn't read from his vantage. Laykin argued something causing DeWald to respond by hammering the white board with the tip of the marker in his hand, obviously trying to enlist the aid of visual emphasis. Thomas returned to his drink when suddenly the door swung open and Laykin trounced through. He jammed some change into a vending machine.

"You guys gonna cool down anytime soon?" Thomas asked.

Laykin bent over and pulled a candy bar from the vending machine. He waved the bar at Thomas as he spoke. "The feds can kiss my butt! He wants me to wait on bringing in the Ackerman kid. Can you believe that crap! He lets the kid vacation at Nantucket—what the hell are we—a travel agency or a law enforcement agency! Hell, let's send the kid to Disney

World, maybe he can strangle one of those whiny voiced kids in the tunnel of *It's a small world after all*."

Thomas grinned. "Sounds like you're a little pissed off."

"Screw this, I'm going home. I've put in enough Sundays." Laykin ripped the wrapper from the candy bar and consumed half of it in one fumed bite. He stomped out the door.

Thomas looked out the glass and observed DeWald still standing in the same place, rubbing his chin, and drawing on the white board. He shook his head and dropped the marker to the tray at the base of the white board. He turned and faced Thomas with a puzzled look. Moments later he had joined Thomas in the break room.

"I need a drink," he said, just inside the door.

Thomas pointed. "Vodka machine is the second one from the end there."

DeWald grabbed a can of Coke and sat at the table with Thomas. He took a couple of sips in silence before speaking up. "Who put the burr under his saddle?"

"He's very dedicated."

"I think there's more to it than that. Could be the pressure coming from above I guess. It's easier on me, I don't work for the Chief."

Thomas smacked his lips. "There's more to it."

DeWald cocked his head. "What might that be, Junior?"

Thomas played with the Mountain Dew can, rotating it on the table. "Laykin started out his law enforcement career in Champaign, Illinois. He made detective not long after. After twenty one years on the force, he was abruptly let go."

"What happened?"

"Laykin has a brother who had three kids. The oldest, Caroline, attended the University of Illinois. During her junior year, there was an outbreak of rapes off campus. Laykin was assigned to the case. He and his colleagues concentrated their efforts on one particular suspect. Apparently they were close to a very solid case, but the DA wanted more before pressing charges. Said there was too much circumstantial and not enough physical. The suspect, a kid on campus, presumably raped three women and it was obvious he would rape again, which he did, but for some reason he added a twist with the last one and decided after raping her to kill her too."

DeWald puffed his cheeks and blew out slowly. "Let me guess...Laykin's niece."

"Precisely. If the DA had gone forward with the case when Laykin wanted to, the suspect would have been off the streets, Caroline would be alive today."

"Just when I thought it couldn't get worse. I've got a real challenge, don't I?"

"More than you can imagine."

DeWald nodded. "But why was he let go?"

Thomas crunched the soda can, folding it in half, and tossed it into the garbage can a few feet away. He got up from the chair and straightened his cloths. "So he could collect himself."

"Huh?"

"After they finally did book the guy, he went to pay him a visit. He told the guard to leave the two of them alone under the auspices that he would be questioning him. The guard heard screaming and returned moments later. The kid was on the floor, badly beaten. Laykin's knuckles were dripping with blood. He beat the crap out of the kid."

"What a loose cannon."

Thomas shrugged. "Kid recovered and had his trial for the murder. The evidence was overwhelming. The jury convicted, spending less than half an hour in deliberation." Thomas reached for the door. "You know, if you ask me, Laykin gave the kid what he had coming to him."

DeWald read the pleasure in Thomas' face. "You're baring your teeth, Junior."

"Scum like that—I've got no patience."

"So is he planning to use those same tactics again?"

Thomas pulled the door open. "Who knows? Funny though…"

"Yeah…"

"I once asked him whether the loss of his job was worth the satisfaction he got out of beating that kid senseless."

"And…"

Thomas shrugged. "He just smiled."

Thomas stepped through the door and closed it behind him, leaving DeWald behind to ponder it all.

The ferry moved in slow motion. The wake rolled slothfully over the seas. The minute hand on his watch was two minutes further along than the last time he looked. The deck, Bennett now had memorized, the number of chairs skirting the railing: 38 in all. The flattened piece of chewing gum shaped like Dick Nixon, the styrofoam coffee cup sequestered tightly under one chair, the cigar butt, which he at first thought to be a disgusting piece of something else. He watched the seas in front of the ferry with a weary eye, the turgid surface banging against the bow of the ship. Each white crested comber delaying progress, he noted, when the ship slowed slightly after each collision.

He prayed Hoosier was okay. As he paced the deck of the ferry, and when he was not thinking over its slow progress, she consumed his thoughts. Somebody had entered his house. Could it have been the cops? Or maybe, odder yet, Alfred? He had practically done it before. But why would Hoosier not call him back? Bennett remained online for over two hours afterwards, waiting, hoping Hoosier might connect up again. He noted she was no longer online thirty minutes after their conversation, consistent with the auto-timeout at Pequod. The auto-timeout at Peqoud would have knocked her offline when it detected no traffic for over one half-hour. He knew her well enough to know she was capable of returning to her net phone, if she could have. She was detained, in some manner—that he was sure of.

During the ride he dedicated a chunk of his time to trying to determine whether he was going to ask Hoosier to drop out of his pursuit. The work was getting decidedly dangerous. He pictured the long line of protestors marching beneath his window at work—and at the courthouse, his paranoia as to whether they falsely thought he was the perpetrator. The detectives were so close behind that he could smell their cheap cologne. It was only going to get worse. Why put Hoosier through it? The implications of a bad outcome were life destroying, if not threatening. If they convicted him, she would be on his heels with an aiding and abetting charge. He tried to think of a good reason for her to remain at his aid, but his thoughts swam in a void. He knew the ramifications—they could no longer spend time together—not even give the appearance they were. He made his decision: she was dropping out. But was she okay?

As he leaned over the railing, eyeing the seas below, he searched for solace over his decision—some peaceful image conjured in the mist, but he came up short, other than to realize that if he got out of the mess, he could picture them together again. It was a big *if.* He made his site tempting enough, he thought, but as tempting as it was, his friend had to find it first. And what if he surfed once a week—or, worse yet, what if he never surfed? Waiting was not something Bennett did well. He preferred to be proactive, and he knew he would be, but he also knew his best chances would come from waiting—a huge disappointment to his normal thought process. Aunt Grace had helped. She had a special way about her, as if a part of God were inside, living and breathing as her soul mate. And when she reached out to comfort him, knowing something was wrong, without him ever telling her, the hand of God joined in, instilling a modicum of peace. When he left her cottage in a rush that morning, telling her only that a friend might be in trouble, she had said a prayer, adding a mysterious calm to his departure.

He glanced at his watch; it was 6 minutes later then the last time he looked.

He jumped from the taxi and dashed across his yard to the front door of his home not knowing whether DeWald had straightened things out with the local police, and not really caring, other than for fear of being tackled from behind before reaching the house. When he arrived at the front door it occurred to him Hoosier had his keys. For the first time he looked over his shoulder thinking he might find pursuit, but a neighbor kid bouncing a basketball in the driveway across the street was the only other person in sight. He forced himself to calm down and to think. He had no hide-a-key, that he knew, leaving him only one option, to break in through a window. He walked back behind the house and looked at his options. He decided on a basement window. The glass was cheap, probably easy to break, and should be easy to replace. He gave it a good kick with his shoe and was startled to see the window stubbornly intact, despite the loud bang that resulted. He tried again with the same result. Frustrated he combed the yard for a rock. He dislodged a softball-sized rock from against the fence and started back towards the window.

"Bennett, what are you doing?" A female voice whispered through the sliding glass door.

The voice startled Bennett, his hand losing its grip on the rock and it thumped to the grass below. The curtains over the sliding glass door hid the voice. "Hoosier?" he asked, hopefully.

The door slid open and out stepped a woman in rich auburn hair. "Hoosier? Who's that?"

A week ago the thought that he would have been disappointed to see Tanya would have been an insane one—but not today. He replied with a pale hello.

She smiled and raised her brow. "Don't act so glad to see me."

Bennett scrunched his face. "How did you get in?"

"The front door," she answered looking surprised, "it was unlocked. I knocked a couple of times, and for whatever reason I tried the door. I guess I was concerned why you weren't answering. I noticed your car was here."

Bennett nodded, pursing his lips. "How long have you been here?"

"Ten, fifteen minutes, max."

He sighed. "Look Tanya, it's nice seeing you and all, but I'm really busy."

"Are you in some kind of trouble?"

"No."

"What's with the rock and all?"

He feigned a chuckle. "I'm so absent minded lately. I went for a walk and forgot to take my keys. I thought I had locked the front door. Breaking a window seemed like the only way in."

Bennett followed Tanya's eyes over to his bag, which lay crumpled by the basement window.

She smiled. "Must have been a long walk."

"Actually, I left that out the other night—"

"You don't owe me any explanations, Bennett." Tanya looked at the ground. "Look…you look dead on your feet, why don't you come inside."

Bennett nodded. He walked over and picked up his bag and joined Tanya at the door. The two of them shuttled into the living room.

"Let me make you some coffee," Tanya said, heading towards the kitchen.

Bennett settled on to the sofa. "That's okay, Tanya—"

"It will only take a few minutes. It's the least I can do for you after the way I've treated you lately."

Bennett groaned as he bent over and removed his shoes. "Everything you need for the coffee is in the side cabinet by the sink. So how did you find where I lived?"

"Internet—I'm using it more and more, thanks to you."

I'll bet you are. He looked over at the phone thinking he needed to call Hoosier. But he didn't want to with Tanya there, what difference it made, he really didn't know, but it did.

"I'm sorry I haven't been in contact lately," Tanya said from the kitchen, the drip coffee machine steaming over the counter where Bennett could see it.

"It's okay, its not like I made the effort myself." Bennett eyed the phone again.

"Maybe that tells us something."

"Yeah, maybe you're right."

"What do you think it tells us?"

It tells me I don't have time for this, Bennett thought. "That we probably are just meant to be friends."

The coffee machine sputtered out in gasps the last drops of water remaining, into the filter.

"Sugar?"

"Black please."

Tanya entered the living room carrying a mug of coffee. She rested it on a coaster setting on top of the lamp table next to where Bennett sat. "Be careful, it's hot." She left the room, returning with a mug in her hand. She sat in the recliner across from Bennett, arched over, the mug cuddled in her palms, allowing the steam to rise up to her face. She lifted her head. "I was kind of thinking the same thing."

"That we should just be friends?"

"Yeah. I hope that doesn't hurt you."

He wasn't sure how it made him feel—confused maybe, but deep inside he couldn't be sure if he even wanted to be her friend. "Don't worry. I'm okay with it. I mean you're beautiful, attractive, but you know...things just didn't click."

Tanya sucked on her upper lip. "Thanks, and I agree, sometimes things don't click."

Bennett nodded and glanced at the phone for the third time. He could see Tanya following his eyes.

"Well, hey...I better be going," she said, getting up from her chair. "I'll see you around. Right?"

Bennett nodded. "Sure, of course."

Tanya walked to the kitchen, rinsed her mug in the sink, and picked up her purse that was resting in the hallway. She paused once she had her hand on the doorknob. "I would kind of like to get to know the other folks in the group over at the NBWHS. So I might be spending some time with your friends. I hope that won't upset you."

Bennett turned to face her. "Nah, don't worry about it. I'd stay away from Elander though, the guy's nuts."

"I'll keep that in—"

The door swung wide, opened from the outside, nearly knocking Tanya to the floor. There, just inside the door wearing a startled look, stood Hoosier.

The spectral upper torso of the woman floated three feet under water, wedged between the jagged lips of two large boulders, illuminated in the dark by a powerful streetlight, orbited by millions of nocturnal bugs. The seaweed that bearded the boulders mocked a grass skirt at midriff on the body. One white arm lay outstretched at the side, the other eerily pointing to the water's surface. And in the wavering current, it swayed, beckoning those who stood at the causeway edge peering down upon her.

A cigarette butt landed at the water's surface to the side of the body, sizzling as it made contact.

"Jeez, Laykin, have some decency," DeWald said from his perched position on the edge of the causeway.

"Need I remind you the woman is dead, DeWald." Laykin lit up another one.

DeWald shook his head, continuing to look down into the water. The sea and its inhabitants were not kind to the soft tissues of the human body. He concentrated on her forehead, attempting to make out lines there that conceivably might define a whale. He rose to his feet. "Does she fit the Sheldon description?"

"Difficult to tell from here wouldn't you say?"

"I see blonde hair, a potential butterfly tattoo on her right breast…seems likely to me."

Laykin shook his head. "You've got far better eyes than me."

"So I guess since he didn't get his whale, he just dumped her in the ocean."

"Maybe our killer isn't so meticulous after all, maybe he breaks away from his doctrine when urgency dictates it."

"What are you saying?"

"I'm saying that if the police were to pick him up, as we did in fact do with Bennett Ackerman, and question him, as in fact we did with Bennett Ackerman, that after his release he might conclude that getting rid of the body was more important then waiting on his whale."

"That's one possibility."

"And the others are…"

"He couldn't get the whale he needed, so he proceeded without it."

"Look DeWald, this guy has gone through extravagant, if not Herculean efforts to ensure each victim is laid to rest in the corpse of a whale—do you really think a delay of a day or two is going to divert him from his rituals?"

"If the victim's rotting body is stinking up his house he might."

Laykin shook his head and puffed a smoke ring into the still night air. "I don't get it, DeWald. What's this boy got on you? Are you two lovers or something?"

DeWald's posture stiffened. "You're out of line Laykin."

Laykin smirked. "You're the Feds, what do I know?"

The two stood in silence, eyeing one another. The streetlight high above, casting short shadows behind them. A spotlight off the back of an emergency vehicle burst on the scene, illuminating a large chunk of the road. Divers carefully released the body from the grip of the stones and it was carried up the side of the causeway to the road at the top. Quickly the body was zipped in a white body bag, just before an approaching car passed by.

DeWald turned away from Laykin, who had not budged, and walked over to the body bag. He unzipped the bag down just enough to study the woman's forehead. There he found the tattoo of a whale, much like the others. He removed his cell phone from his pocket and placed a call.

"Who'd you call?" Laykin asked, after DeWald had returned the phone to his pocket.

DeWald didn't feel much like answering, but he was a professional and he would deal with Laykin professionally. "Our computer forensic guy, I want him taking apart Ms. Sheldon's computer tomorrow."

"How do you know she has one?"

"I just know she does."

Bennett ran to the door, throwing his arms around Hoosier. She gasped from his zealous hug. Tanya looked on, nodding, acknowledging the undeniable affection Bennett held for Hoosier. In silence, she made her way behind Hoosier to head out the door. Bennett didn't notice.

"Wait a minute," Hoosier said, talking over her shoulder while observing Tanya.

Tanya stopped. "I really need to be going...I was just checking in on him—I was worried that he might not be okay."

Hoosier shook her head. "Why?"

"Why what?"

"Why were you worried about him?"

Bennett released his hold on Hoosier and stepped back a few paces. "I hadn't called her in a while. We needed to sort some things out concerning us."

Hoosier turned pale. "Oh, yes...how stupid of me, I had forgotten. I'm the one who should be leaving—"

Hoosier's change in color did not go unnoticed by Bennett. "We're just friends. We made a mutual decision to be friends. Both of us agree we're not cut out to be anything more than that."

Hoosier turned her gaze on Tanya, who nodded. "Just one of those things that didn't work out," she stated matter-of-factly, and walked out the door to her car.

"Are you sure about this, Bennett?"

Bennett smiled. "Come here..." He placed another large hug on her. "You know how I suspect her of being involved. Man, I was so worried about you."

"Why?"

"The other night—your last words were 'someone's in the house'."

"Oh, man, I guess you never received my last message. I told you it was just the cops. They found the front door unlocked—or so they say—and felt obligated to investigate if you were in any danger."

"Yeah, right," Bennett said, sarcastically.

Hoosier scanned Bennett as he stood before her. "You're not looking too good."

"Thanks a lot. It was a short night and long trip back."

Hoosier smiled and said, "Because of me?" Bennett nodded. She gave him a kiss and said, "you're so special! Now let's get to work."

Bennett popped his eyes. "Huh?"

"The clock is ticking…"

"I thought I might get a short break."

"Tell me what you've learned."

The two went to the living room where Bennett took a sip from his tepid coffee. He went over the web site, going into detail as to how he set things up. "If I'm in prison, you're going to have to know how to work this thing." He watched Hoosier's jaw drop and tried providing her what encouragement he could drum up at the moment. He finished by telling the tale he had heard from the fisherman at the pub.

"Bizarre…" Hoosier said, blankly shaking her head. "Do you think there's some kind of link there?"

"Definitely."

"So we're looking for the one-armed man?" Hoosier grinned.

Bennett sighed. "Cute, Hoosier…real cute. But worse yet, he, the killer I mean, might be dead and buried at the bottom of the ocean. How will I ever prove my innocence if I can't produce the real killer?"

"He's still alive."

"How do you know?"

"I don't. But I sense it."

Bennett got up from his chair and motioned to Hoosier to do the same. He headed upstairs with Hoosier in tow. In his office, he had Hoosier pack up her laptop and he unpacked his borrowed laptop and connected with Pequod, telling Hoosier he would show her the web site. He decided to check his mail first and found another letter from Griggs—a waste, probably best just to delete it.

"What are you doing?" Hoosier asked, as the letter was moved to a trash folder.

"The guy is wasting my time."

"But it says it's important."

Bennett could sense Hoosier would insist he read the mail. Rolling his eyes, he opened the trash folder and opened Griggs mail.

Hey, I'm finally learning some Unix! Anyway, I checked my system log files for the 24-hour block, during which time the whale file arrived on my server, and found an interesting entry. It says:

"A treasure chest has been buried on your server somewhere!"

This is the only unusual entry I found. Appears to have no correlation, but you never know. Perhaps it is useful to you in some way.

Later, Griggs

Bennett caressed his chin, intrigued that the odd message was a contemporary of the whale file, but at the same time, confused by its seemingly unrelated content. It was almost as if the two were in sync time wise, but completely out of sync content wise...why would that be? He picked up his phone and called operations at Pequod. When Adam answered he wasted no time in asking Adam to search the system logs, covering the last month, for the word *treasure*. Adam questioned the sanity of the request at first, but after Bennett made it clear it was important, he said would get right on it, knowing it best not to second guess his Chief Engineer.

"Can you make any sense of it?" Hoosier asked.

Bennett swished his lips. "On the surface no, but there's something there, something I can't grasp at the moment, it's just beyond my reach, but it's there. Isn't that strange?"

"Maybe it will come to you."

Bennett swallowed hard. It escaped him up until now, but the decision he made on the ferry ride home, came back to him. He turned in his chair to face Hoosier. He could see expectation in her eyes. "Now, Hoosier, let's talk about your involvement in this..."

A lone seagull, perched atop the brass-crested cupola of the New Bedford Whaling Museum, jostled the strong winds gusting from the bay to remain in prime position for overlooking it. Bennett watched, thinking that at any moment the wind would claim its rightful position as king of the mountain. Pinched beneath his elbow, against the glass top of the patio table, was a copy of the daily newspaper. The pole, supporting the umbrella over his head, thrashed around inside a hole in the middle of the table. He clung to a cup of gourmet coffee. The outdoor court of the Gourmet Coffee Shoppe at William and Melville was empty. Despite the strong wind, Bennett yearned for the fresh air and he stubbornly remained a fixture out front.

Convincing Hoosier to drop out of the pursuit for his freedom had proven futile. No argument he presented, regardless of merit, dented her fortitude to press on. She told him there would be no aiding and abetting because there would be no trial—let alone an arrest—she was sure of it. And when he argued, hypothetically, that should there be one and should he be convicted, she too would probably go to prison, she responded that accepting that risk was her decision to make and not his. In fifteen minutes the discussion was over. Hoosier would remain in the pursuit.

Bennett chuckled as he thought over the exchange. He never stood a chance at winning it, that he knew, but also he knew, deep inside, he was glad it was that way.

"You always sit out in the wind like this?"

Bennett looked up to see DeWald tableside. "Only when I've been falsely accused."

"So, this is your fist time sitting in the wind?"

Bennett nodded to the chair beside him and DeWald took a seat. "I wanted to get caught-up," DeWald said.

Bennett sniffled. "Look DeWald, to be perfectly frank, if you expect me to trust you and provide you information, shouldn't I be expecting the same from you?"

"There is a difference, you have no choice, but I do."

Bennett shook his head, not hiding his disgust.

"But..." DeWald continued, "I want this case solved. I don't want another innocent woman to die. That makes me, in one sense, desperate. And the honest truth, I think you've got every bit as good a chance of solving this thing as I do."

Bennett pursed his lips and nodded. "So what can you tell me?"

"I just got a report early this morning from our computer forensic expert that the last victim's computer had saved transcripts from chat sessions with a certain individual. That same individual appears in chat transcripts on all the other victim's computers as well, with the exception of Ms. Rueger."

"Who is it?"

"The same individual that's been playing games with you—Mocha Dick."

Bennett nodded slowly. "So this Mocha Dick is the common thread through all of this. That pretty much confirms the conclusions I was drawing."

"You track him down, this case might be over."

"I'm working on it. But this is more than a human being."

"What are you saying?"

"I'm saying, as you are assuming, there is definitely a human identity behind the Mocha Dick name, that's the voice we're seeing in the chat room transcripts and other places. It's just a hunch, but I believe there exists another component to Mocha Dick, a bits and bytes side—a rogue process in cyberspace. At least the voice alludes to something like it."

"So if this digital beast exists, what would be its purpose?"

"Undetermined. I have some ideas, but once again, we're talking mere hunches."

"Well I wouldn't waste my time on it. It's the voice we're looking for, the one in the chat rooms. We can't exactly toss some digital beast in the slammer."

"I disagree, if a killer owned a dog as a pet and that pet knew its way home or had its home address on its collar, would you pursue it?"

"Touché. So a cyber dog."

Bennett rolled his eyes. "Please. Enough of this cyber prefix on everything. Let's leave it at cyberspace."

DeWald nodded. "Fine. So tell me, what are you doing?"

"Everything I know how: setting a trap, sending agents out into cyberspace to try to pick up his trail, in communication with a guy on the west coast who has found some evidence—"

"Wait a minute." DeWald cocked his head. "Why would there be evidence on the west coast. You're not finding evidence of crimes there as well, are you?"

"No, but you have to lose this physically geared mindset of yours. This evidence can traverse the globe at the speed of sound, and in most places, the speed of light. To move files to the west coast is like picking up a glass in your living room and moving it into the kitchen. Where he puts the evidence is not an issue, why he puts it there is."

DeWald nodded. "Makes sense. So why would he put it there?"

"I don't know. I thought maybe to hide it, but then why would he put it out at all? It must be a safe haven for whatever it is he's got going. Hackers do it all the time. They keep their tools of the trade on some unprotected server, where the administrator is asleep and security is weak, and then launch their attacks from there."

"Kind of like launching an attack on a country from an aircraft carrier out at sea."

"Close, only the attacking country is using some other country's carrier to attack from, and the artillery is marked as belonging to the country with the carrier."

"So this guy is leaving nothing around that is marked New Bedford?"

"Not likely. With the exception of the occasional *whale* file he likes to leave behind at Pequod. But I don't believe our friend is really *attacking* per se, I believe he is using electronic real estate to play out his fantasies. I honestly think it's a game of sorts to him. Kind of like Dungeons and Dragons under a Moby Dick cloak, and somehow Mocha Dick is a part of it all." Bennett drifted back in his chair. "You realize this guy could be living anywhere in the world, don't you?"

"Could be, but isn't."

"He could be remotely directing some local guys from around here to do his dirty work."

DeWald shook his head. "Let's move back to the physical realm where my expertise lies. In almost every case where you have a serial killer operating, he has a strong desire to physically commit the crime. That's his drug, it's what gets him high, a long distance crime to him, would be like taking a placebo and knowing it to be one. There's no thrill in it. I can almost guarantee the guy is here locally." DeWald grabbed a sugar packet and rotated it with his fingers between his two hands. "Are you ready to tell me where you got those images."

"Images?"

"The ones you said we would find on your computer—the ones that aren't yours…"

Bennett took a sip of his coffee and swallowed hard. "They're not relevant at this time."

"How can you say that, whomever had them must have some link to the killer, or better yet, be the killer himself."

"Could this killer be a female?"

"Interesting thought, but no, not unless she were extremely strong. Why do you ask? Did you find them on Hoosier's computer?"

Bennett's face whitened. "No! Keep her out of this."

DeWald cocked his head. "Did I touch a nerve?"

"Look she's literally risking her life to help me out. I want you to understand that. No matter what happens, she is just trying to help me out. That's all. If I get falsely accused, I don't want her brought down with me."

"You know there are no guarantees." DeWald tossed the sugar packet towards the hole in the center of the table. It bounced off the umbrella pole. "We checked into Alfred Huffnagle and it led us nowhere."

"He sure seemed to be around my place all the time, bugging the heck out of me."

"Do you really think he has the computer acumen to pull it off?"

"Have you seen programmers, I mean good ones? They're the strangest lot on earth."

"He's just a lonely old man."

"He's got a wife."

DeWald shook his head. "He made all that up. He's alone with no friends, and that's why he showed up one night at your meeting. He was looking for friends."

Bennett felt a pang of guilt, but chose not to display it.

"Now for some advice," DeWald said seriously, "keep records of everything you do. What evidence you discover, whether electronic or physical, make sure you document and keep in a safe place. Keep copies and give them to me if you like. When this comes to trial, your records may become vital to proving the state's case."

"I know the game I'm in."

"Take it from an experienced law professional, you can't afford to lose any evidence." DeWald rose from the table. "And Bennett, if you catch this guy, no matter who it may turn out to be, you call me immediately and turn what you have over to the authorities."

"Of course I will. Why wouldn't I?"

"You won't tell me where you got the images, why would this be any different?"

"Why do you do this?"

"Do what?"

"Do what you do. You're out of the Boston field office? You must be away from your wife and family for weeks at a time."

"I see them on weekends. They understand."

"You must really be driven."

"Aren't we all?"

"Hypothetically, if you knew your wife was gonna die next week, would you be here?"

"Of course not."

"You never know when they're gonna die, DeWald."

"What are you saying? Is there something you're not telling me? Is my wife in harm's way?"

"No, nothing like that. But I know what it's like to lose someone—you can't get the time back."

"Well let's get this damn thing closed—I can home and you can get on with your life."

Bennett watched as DeWald walked around the corner of the coffee shop. Before leaving, he glanced up at the cupola and found the seagull still perched and still fighting the wind. Hang in there my friend—we're in for the ride of our lives.

The report was lying in his chair when Bennett arrived at the office. He immediately recognized it as a series of entries gleaned from Pequod's system logs. In each instance, Adam had highlighted the word *treasure*. He cross-checked the time and date stamps he had documented for each of the *whale* files that had been found on Pequod with the time and date stamps for the log entries: they were within milliseconds of each other. He had to conclude the process that delivered the whale files produced the side effect of writing an entry directly into the system log files. How sloppy. But the two seemed unrelated—so why do it? Bennett reached for a pencil to jot notes on the report when he noticed somebody standing in his doorway. He looked up to see the large figure of Collin leaning against a side of the doorframe.

"Collin...you're quiet today."

Collin gave a minute shrug. "Hey. Didn't want to disturb you."

"You know you don't have to worry about that."

"What're you reading?"

Bennett explained what he discovered. Collin cocked his head when Bennett mentioned the log entry. "Makes me think of Long John Silver's— what are you doing for lunch?"

Bennett steepled his fingers and brought them to his chin. "Usually hackers remove entries from logs, they don't add them."

Collin nodded. "Anyway, I just wanted to say I'm sorry about flying off the handle the other day. I was out of line. I just don't like the Feds in our britches...not unless it's absolutely necessary." He moved to the seat across Bennett's desk and slid forward until his knees met the desk. He leaned forward, placing his elbows on the desktop, held his arms wide and dropped his voice. "We can handle this guy ourselves. Do you really think they've got better people than we do? We're the best." He pointed to Bennett and then to himself. "We don't need outsiders. If this guy is getting out of line, then maybe its time to trash him. I'll be the first to admit that my plan was to ignore him. Usually these morons are looking for attention— you ignore them—they go away." He sighed and leaned back in the chair. "But obviously this guy's got you chugged full...and as a friend I should be supporting you and not fighting you." His eyes dropped to the floor. "Let's catch this moron—I mean an all out assault—we'll team, you, me, Will, Adam...whoever we need, whatever it takes."

Bennett sat silent for a moment. "There's more to it now, Collin."

Collin knotted his brow. "What do you mean?"

Bennett shook his head. "I really can't tell you, other than the stakes are higher now."

Collin leaned forward. "Bennett—we're good friends. What can you possibly not tell me...come on, it's Collin you're talking to."

"It's for your own good."

Collin pursed his lips and stared out the office window. "I see."

"You see what?"

"You've still got your feathers ruffled."

"No, It's not that—"

"Then what in the world is it?"

"It's something that I'm close to resolving myself. I don't need a team."

Collin rose from the chair. "I don't know which one it is, Tanya or Hoosier, but one of them is screwing with your mind. I guess the old crew at Pequod and the NBWHS isn't good enough for you anymore. You've gone back on your raisin."

Bennett scrunched his face. "My what?" He shook his head. "You're way off base."

"Oh yeah, then why haven't you been around the NBWHS, or Uncas, or any of the places we hang out?"

Collin stomped from Bennett's office. Bennett walked over to the window and looked down at the street below: no protestors were marching. It was for the best that Collin and the other guys stay out of it. It was bad enough that he might bring Hoosier along to the state pen, but it would be worse yet to bring a whole group along. There was a time when they might have helped, but he knew that time had past, he knew from this point forward to the home stretch, whatever that might be, it was he and Hoosier and nobody else.

Hoosier's car was parked at the side of The Skipper, a well-known stop on Sconticut Neck road. She phoned Bennett that morning suggesting they meet for lunch. She wouldn't say why, but even if she had tried, Bennett wouldn't have heard it, children were screaming so noisily in the background on Hoosier's end. He parked next to Hoosier's car and got out to a strong whiff of hamburgers frying on the grill. The Skipper never managed to put in place air conditioning, and during most days it wasn't needed, today included. A screen door was all that kept the flies out. The owner never put anything more in place realizing that the smell of his grill wafting down the street proved to be fantastic advertising.

Bennett found Hoosier at a booth. He slid in across from her. They exchanged smiles.

"Sorry to drag you away from work," Hoosier said teasingly.

"Please...you know there's nowhere else I'd rather be."

A waitress stepped tableside and took their orders. Hoosier waited until she left before speaking. "Tanya stopped by this morning."

"Tanya?" Bennett said incredulously.

"Can you believe it?"

"Why?"

"Interesting question. There's something about her I just can't seem to put my finger on...Anyway, she said she just wanted to make sure I understood the relationship she had with you. That I should not perceive her as a threat."

"What did you say to her?"

"I told her not to worry about it. That you were a big boy and I was a big girl and it wasn't an issue, really."

Bennett nodded. "Strange..."

"So anyway, that was the first five minutes of our conversation."

"There was more?"

"Yeah, she started asking me questions about myself."

"Like what?"

"Standard stuff: where I was from, how did I meet you, was I into the whaling stuff, and so on."

"I wonder why she cares?"

"Maybe wants to know the enemy." Hoosier grinned.

"I guess that could be."

"No, I don't think so because she proceeded to ask me questions about your friends and our friends."

"Really?"

"Yeah, Elander, Adam, Collin, Will, Patrick—"

"Patrick?"

"Yep."

"Hmm. She's just looking for all sorts of information."

The food arrived and the two ate and pondered in silence for a moment. Hoosier took a paper napkin to her lips and said, "Let's assume something for a moment. Let's assume she is somehow involved in our situation. Why ask questions? How would it benefit her?"

Bennett stopped chewing and thought for a moment. "I guess the information might provide her the additional means for framing me on this thing."

"Do you really think she's capable of that type of behavior?"

"Who knows? There's the thing with the files."

"But you took them, she didn't plant them on you. I think it's something else."

"I can't figure out how she fits into this mess. What else is there?"

"Not sure. I wish there were a way to find out."

"I could do some 'research' on her. Just a hack or two might illuminate some things for us."

"Nah, we can't risk you getting caught."

Bennett laughed. "Me? Surely you jest."

"I know you wouldn't, but, if on a long shot you did, it would be all over. And besides, there's a quicker more efficient way, an old fashion way..."

"And that is?"

Hoosier shrugged. "Stake out her house, look through her trash..."

The verdigris frog popped into Bennett's mind. "We could go one step further."

"How so?"

"I know how to get in."

Hoosier lowered her voice. "Breaking and entering?"

"Technically no, just entering."

Hoosier shook her head. "Same result, you get caught, it's over."

"But it would probably answer our questions immediately."

"I think it's too risky…" Hoosier's brow furrowed. "However, it may not be too risky for me."

"Out of the question," Bennett said sternly.

"Where does she live?"

"It's not a good idea for you to go."

"And it is for you? They'll follow you there and arrest you on the spot."

"I can lose them."

"Maybe. But nobody is following me. Look, the worst that can happen is she stumbles in on me. I can simply tell her I found the door unlocked, you know, just like she did with your door. I'll bring a cheap gift along with me. I'll tell her I just wanted to bury the hatchet."

"The hatchet's been buried ten times already."

"So, I'm a little thick, that's not a crime."

"I don't know…"

"The address?"

Much to his better judgement, Bennett found himself jotting down Tanya's address on a napkin and handing it to Hoosier. Severe circumstances called for severe action. Earlier, on that long ferryboat ride, he came to realize he was playing on a different field now. The rules of the game were much more lax, the line between black and white, dithered to a gray. Survival dictated violation of the lesser laws. When one faces possible conviction for breaking the ultimate law, the others all pale, and the consequences of their transgression pale with them. In a warped sort of way it felt liberating. After all, if you've supposedly murdered someone, shoplifting a candy bar from a 7-Eleven, on a relative scale of transgressions, is tantamount to an innocent act. Bennett looked up at Hoosier who had finished her lunch. "When are you going over there?"

"A little later this afternoon."

Bennett squeezed her hand. "Be careful."

"I'll let you know as soon as I'm out of there."

They kissed and Hoosier stepped into her car. As Bennett approached Pequod, he prayed he had not set something into motion that he would regret the rest of his life.

247

DeWald put down the autopsy report on the Sheldon girl and took a bite of his crabcake sandwich. The findings matched with the other three killings: the cause of death was strangulation—fishing line used for a ligature—she had a whale tattoo on her forehead, and chemical traces of perfume were found in her hair. And, once again, the killer left none of his own physical evidence behind, trace or otherwise. DeWald thought of Bennett, hoping his traps would successfully snare the killer. He hated to admit it, but he knew with each passing day, he was relying more and more on Bennett to break something free. It wasn't good investigative technique, and it chafed his ego knowing that someone else had the upper hand on solving the crime, but he knew he could not allow that to matter. He would continue to do his best by tracking down any leads as they arose, by interviewing relatives or owners of shops where the killer might have been a patron, even taking guesses if he had to, he would do it all. But he knew his best bet was Bennett.

His pager vibrated at his side. The display indicated Thomas' number. Thomas picked up after a single ring.

"I shouldn't be telling you this," Thomas said, without even waiting for DeWald to say anything.

"What is it, Junior?" DeWald's eyes moved rapidly around the restaurant.

"Laykin is meeting with Wally Atkinson."

"Damn him!"

He burned rubber out of the restaurant parking lot, causing everyone with window seats to stare outside at the commotion. He didn't care. Laykin was with the DA, pleading with the man to take action, to make him see that there was enough evidence to make an arrest and build a successful case. Though he really didn't know Wally Atkinson, he knew what pressure could do to a DA, how much public opinion meant, and how an arrest would put ballots in the box. Timing was everything, if the DA waited too long, the public might conclude he was not performing his duties. If word got out that the police had a case and he failed to take action, well…DeWald screeched into a no parking zone in front of the city building and clambered up the front steps. Completely out of breath, he flashed his badge at the secretary, and barged into the DA's office.

Seated in a chair in front of the DA's desk, Laykin jerked his neck around. The DA too, snapped his head in DeWald's direction.

"I believe I should be part of this meeting, gentlemen," DeWald said, taking a seat in a chair next to Laykin.

"Agent DeWald, right?" the DA asked.

Out of breath, DeWald nodded and glanced at Laykin. The man returned a stern glance.

"Well Mr. DeWald, I believe this is a city matter."

"I understand that Atkinson—"

"Call me Wally, please…"

"Wally, I understand that, but I'm an integral member of the investigative team. My opinions should be heard, FBI or otherwise."

"Wally," Laykin said, leaning forward in his chair. "He plays a support role in this investigation. Sure, we asked for assistance from the FBI and agent DeWald has been of some assistance. But he and the FBI are not part of the decision-making process for this city, nor should they be. This is a city matter, not a federal matter. He can formulate opinions regarding the evidence, but he shouldn't be a part of the process that formulates opinions about the course of action this city should take."

Wally nodded. "Well spoken Detective Laykin. However, I must assume, given that Mr. DeWald apparently busted his butt to make this meeting, that he has some sort of a dissenting opinion on this matter. I can't imagine him barging in like he did if he were in agreement with you."

"I'm not sure what Detective Laykin has said to you up to this point, but I want to make sure we don't make a major mistake today."

"And what would that be…"

"Making a premature arrest of Detective Laykin's prime suspect."

"It sounds to me like we have a fairly solid case. Two cousins, one a bar owner with a money hungry wife, and a new automobile, harpooning whales and being paid by the other cousin, the wacko, the one doing the terrible things to young women in this city. I'd say his take on things, fits. Do we have to go down the list of evidence in this case line by line to lessen your discomfort with what we are saying."

DeWald shook his head. "That's not necessary, I'm very familiar with the evidence. We have absolutely zero evidence on the cousin, and as for Laykin's suspect, let's pretend we're in a courtroom for a moment—"

"But we're not."

"If you have a reasonable doubt, don't you think the jury might as well?"

"But at this juncture I don't."

"Well, maybe if I told you three out of the four individuals working this case, do—does that change your mind? Or that we have no less than two others who are now strong suspects."

Laykin turned crimson and erupted. "That's bullcrap!" he hissed.

"Calm down," Wally directed.

"Wally, one detective's in love with the boy for crissakes. And DeWald—he's so anal about this he'd need to have witnessed the crime to be convinced, and Thomas, he's with me, so that's a flat out lie!"

The DA held up his hand. "Detective Laykin, settle down."

"I'm the only one on this case with any sense—"

"Enough!"

The room fell to silence, all three men trying to collect themselves.

The DA tapped his fingers from each hand against each other, forming a steeple. He looked over at DeWald. "Sounds like a little prejudice has crept into our team. This detective, has he or she formed an attachment with the suspect?"

"It's a she, Wally, and no there's not an attachment there, but there is respect. She's been working undercover with the 'suspect' and doesn't think he's capable. And that's her unbiased opinion."

Laykin smiled and shook his head. "Bundy would have had her wrapped around his finger like that." Laykin snapped his fingers.

"Doubtful, she's a professional," DeWald said.

"You can't be a professional and have a heart," Laykin said.

"No, *you* can't be a professional and have a heart—" DeWald replied.

The DA slammed a desk drawer and startled DeWald and Laykin. "And Detective Thomas?"

DeWald recovered. "At first he was in agreement with Laykin, and for all I know, he may still be, but…but he thinks it's premature to make an arrest."

"And you?"

"I think our 'suspect' has been framed and I believe there will soon be ample evidence to support my claim."

"So you don't think he did it all?"

"No, I'm almost certain of it."

"Are you certain your thinking is not skewed by what appears to be a 'likeable' person?"

"Of course not."

Laykin laid his hand on Wally's desk. "Need I remind you Wally, we found his boot at a crime scene, his house is littered with psychotic whaling paraphernalia, he has no alibis for the times when the crimes were perpetrated, he suffered a severe head injury as a child, the trauma could be responsible for triggering his tirades of death—"

"You're taking big license with the evidence, don't you think?" DeWald chuckled. "He passed a lie-detector test—"

"He's got TLE for crissakes!"

"Detective Laykin, I'm troubled by Agent DeWald's statements that the rest of the team is not onboard with you."

"So what are you saying? If it takes another woman or two to convince the other three members of this team, then so be it?"

"Detective Laykin, you're not helping your case." Wally looked to DeWald. "What assurances can you give me that Detective Laykin's suspect will not be committing any crimes while remaining at large."

"He's under surveillance, 24/7."

"And I take it this last murder took place before he went under surveillance?"

"It did," Laykin jumped in. "Quite the coincidence. Nothing since."

DeWald rose from his chair. "And I want to add, Wally, that our killer, whomever he may be, is carrying on significant activity on the Internet, a domain this police department and even the FBI at times, is somewhat lost in. Detective Laykin's suspect, Bennett Ackerman, is an expert on the Internet; he is working diligently to track our man down. I think there is a good possibility he may catch the guy. I don't want the opportunity to pass us by. And if Bennett Ackerman is in prison, it surely will."

"Interesting," Wally said, nodding slowly. "That very fact can also lend credence to Detective Laykin's assumptions. If he's good at what he does and you're poor at it, how do you know what he is doing is not one big charade, making a buffoon of all of us? The time we give him may be the time he needs to remove evidence."

DeWald looked at the floor. "I base my opinion on years of experience working with criminals. I know he is not a criminal."

"That's no way to run an investigation Agent DeWald, I don't think I need to tell you that. You've got one week to bring me the preponderance of evidence that tips the scales in your direction and away from Detective Laykin's. And although I'm not in authority over either of you, I would admonish you to work as a team."

"A week may not be enough time," DeWald pleaded.

"You're making a mistake, Wally," Laykin said defiantly.

"See me in one week gentlemen."

The dog, painted on seven ocean—rounded rocks, glued with a powerful cement adhesive, rested in the passenger seat of Hoosier's car. One of the children at the Powerhouse pulled the seven rocks together to make four legs, a body, a tail and a head. The creations were being sold for a fundraiser, so when Hoosier grabbed one from the table out in front of the Powerhouse, laying down a ten, she didn't feel bad over taking it. If Tanya were to surprise her while she was in the house, her explanation of "finding" an open door, and her possession of the dog would make an excellent excuse, she telling Tanya that she was in the neighborhood and had stopped by to see if Tanya would like to make a donation.

Not having navigated New Bedford for some time, Hoosier struggled with the directions, but eventually, through trial and error, arrived at Tanya's home. An empty driveway greeted her, causing her some sense of relief. She grabbed the rock dog and stepped from her car, trying as nonchalantly as possible to comb the neighborhood. Other than a couple of boys who were mowing a lawn two houses down and a woman pushing a baby stroller off in the distance, Hoosier appeared to be by herself. She casually walked to the front door of the home. The two large evergreens at either side of the small front porch provided excellent coverage, only someone directly behind could actually see her. She knocked three times, sensing the beating of her heart, hoping nobody would answer or her little escapade would end abruptly without having accomplished anything. She waited for what seemed like an eternity and concluded nobody was home. She looked to her feet and around the perimeter of the porch. Off in a corner, exactly where Bennett had said it would be, a verdigris frog sat beneath an awning of pine branches. Hoosier bent over and pulled the frog from its cover. She tipped its nose against the palm of her hand and began a series of jarring movements with her arm until the key came flying into her hand.

Here goes nothing. The door opened with a creak and then it occurred to her: what if Tanya had the house alarmed, worse yet, a silent alarm? She was no expert, but she had been in enough houses to know that if the house were alarmed, there would be a control panel located on a wall somewhere. A glance along the walls of the house did not discover any such panel, so she proceeded, knowing the possibility was still very much there. She returned the key to the frog and headed first for Tanya's office, Bennett had told her where it was located. She found the computer immediately, but didn't start it up, knowing she lacked the skills to find anything on it. She did rummage through the floppy disks, scanning the labels for anything of interest, but found nothing unusual. She began pulling out desk drawers and flipping through documents and artifacts located in them. She found a thick stack of papers concerning adoption and foster care. Pages were highlighted, their margins scribbled with notes. She flipped through them and determined the papers were not worth pursuing.

She bent up straight after the bottom drawer, pausing for a moment to think. She already felt like she had been inside for hours and could physically feel time pressures beginning to mount. She contemplated whether to continue her brute force approach, looking through everything, or try an intelligent approach, looking though only those places where something might be hidden. She opted for the latter and headed for the bedroom. Before she arrived at the bedroom door, she noticed what appeared to be an entrance to the attic directly above her head. It was a plain panel embedded in the stucco ceiling. Finding no handle for the panel she went back to the office, returning with a desk chair. She stood on the chair, and carefully applying pressure, tried to raise one end when suddenly it flew up in the air and when it reached a quarter turn, dropped an unfolding wooden ladder from its topside, onto Hoosier's chest, knocking her to the floor.

She lay there stunned by the violent action. It scared her more than it hurt her. She took deep breaths in an attempt to regain the composure she just lost. As she lay there looking up at the ceiling, the way everything worked now made sense. Tip the far edge to cause it to swing up and grab the ladder from the exposed topside allowing it to gently unfold to the floor. Live and learn. She rose to her feet and walked to the base of the ladder, looking up into the attic from there and trying to spot a light switch somewhere in the vicinity. She found none, but obviously she could see into the attic, so there had to be windows up there. The ladder groaned beneath her feet as she cautiously made her way to the top. When her head cleared the floor of the attic she felt a tremendous wave of heat, the sun-backed air inside toasting her body like popcorn in an air popper.

When she stepped on the attic floor, she realized that if Tanya were to show up at the moment, there would be no good explanation for her being in the current location. Checking for termites? Yeah right! A single window in the wall directly in front of her provided illumination for her surroundings. The ceiling and its protruding, menacing shingle nails, only inches from her head, did not appear to contain any light fixtures. She turned her attention to a few boxes resting between exposed joists. She stretched from the plywood platform she stood on to reach the closest box, pulling it to her and onto the platform. It barely wore a coat of dust and had no markings. A lone strip of box sealing tape closed the box tight. Great, Hoosier thought, if she was to open it Tanya would know. Should have known to bring some box sealing tape, she kidded herself. She tried lifting a corner on the tape to allow her to peel it back as a whole, but the corner merely tore from the rest of the strip. Oh well, here it goes…She pulled a car key from her pocket and ripped it through the center of the tape. The flaps on the box sprung free. She pulled them back and inspected the inside. The first item found, a framed glass picture, she could not make out. She carried the picture over by the window. A drop of her sweat splattered on the glass when she reached it. It was a certificate from the Boston Police Academy stating that Stacey Allbright had successfully graduated from the Academy.

Hoosier furrowed her brow, wondering whom Stacey Allbright was and what she had to do with Tanya. A married sister maybe? She crabbed back the box, the small effort causing her clothes to feel damp with her perspiration. She commenced another session of rummaging through the box, this time trying to be more discriminating, not wanting to return to the window unless she was sure she had something of interest. She felt in her hand a thin plastic card. The first thing that came to her mind was that maybe she had an ID card. That was worth the sweaty trip to the window.

She got to the window and paused, trying to catch her breath in the stifling heat. She turned the card towards the window. The card had Tanya's picture, shorter hair and younger, but definitely her. In print was the name Stacey Allbright and information indicating the card was a Boston Police Department identification card. She's a cop! Hoosier sank to the attic floor. It explained a lot, almost everything. She began to contemplate whether further searching was necessary, when she felt a slight percussion coming from the window, she immediately recognized it as a car door shutting, and it had to have come from just outside the window…somebody was there.

She ran back to the box, threw the card in, and slid the box back to its original location—the flaps standing erect—no time to do any more with it. Her legs shot down the ladder, skipping every other rung, and jumping to the floor. She tried to fold the ladder against the pivoting panel in the ceiling, but it kept tumbling to the floor. It occurred to her to fold the ladder up and hold it with one hand while using the other to swing the panel back up into the ceiling. When she tried it, it worked. She shuttled the chair back to the office and flew down the stairs to the front entryway. As her feet made contact with the stone floor, the door opened and in walked Tanya, groceries in her arms. She dropped them to the floor and reached for Hoosier, grabbing her, and seconds later, pushing her to the floor, pinning her against the cold stone floor.

"Who are you?" Tanya shouted, authoritatively.

Hoosier tried to yell out her name, but found it difficult with half her face smashed against the floor.

"I asked for a name!" Tanya repeated.

Hoosier rolled her face from the floor and yelled as loud as she could her name.

"Hoosier?" Tanya said, her voice slightly relaxing.

"Yes, yes, it's me."

Tanya relaxed the hold she had on Hoosier and allowed her to gradually come to her feet. "What in the heck are you doing in my house? And why are you so wet with sweat?"

Hoosier did her best to collect herself. She twisted her head and neck to make sure everything was still intact. "Whew! What was that all about?"

Tanya held her hand to her forehead. "It was purely a reaction on my part. I've been trained...It's automatic for me."

"Where did you learn your technique?"

Tanya pursed her lips. "I'll be the one asking the questions. First and foremost, what are you doing in my house?"

"I've been working the neighborhood, selling some of the products the kids made at the Powerhouse, you know...a fundraiser."

"I see. So what do you have that you want to sell to me?"

"I have a dog left."

"Where is it?"

Good question—where did I leave it? "I left it upstairs in your office—"

"Why go up there?"

Hoosier shrugged. "I—"

"Never mind. How did you get in?"

"The door was unlocked."

Tanya caressed her bottom lip between her thumb and pointing finger. "So tell me Hoosier, as you've gone door to door selling things, who lives next door, they're home all the time, surely you met them."

"I haven't been over there."

"Over where? I don't recall saying which next door. But you're answering my questions. One more question: does Bennett know you're here?"

"I came here first that's why I haven't seen the neighbors yet, and no, Bennett doesn't know I was stopping by. This was all done on a whim. I just thought you might want to make a charitable contribution."

Tanya ran her eyes around the house, including up the stairs to the hallway on the second floor. "So, did you find anything?"

"What do you mean?"

"Don't play games with me, Hoosier. This is life or death for me. If you found something I need to know, and I need to know it now. So, let me ask you one more time," Tanya looked directly into Hoosier's eyes. "Did you find something?"

Hoosier didn't blink. "I don't know what you're talking about, but I guess, given the fact that I was in here for a very short period of time, and given the fact that I wasn't looking for anything, and given the fact that I see nothing unusual here, my answer would have to be no."

"I hope you're telling me the truth, for your sake and mine."

As Hoosier seated herself in her car, wiping the sweat from her brow, she sighed, relieved that the dog actually had been left up in Tanya's office.

Despite the wind, the neighborhood felt still and his house silent. A pervasive sense of emptiness ironically filled the cluttered living room where Bennett sat on the sofa, a photo album spread across his legs. He found himself riding a wave of nostalgia, pushed along by the winds of loneliness sweeping through his day. The realization of the serious nature of his situation had subconsciously triggered a self-imposed quarantine. He was a leper, dangerous to be around, his potential guilt rubbing off on those who sought his company. He hoped Hoosier had built up immunity.

It also occurred to him that it was the day.

The anniversary day.

The day he lost his father—the day that had not left his side for 14 years, the outcome of which he could never forget, head injury or no head injury.

He flipped a page in the photo album. Four more photos appeared. In the upper left quadrant, he and his father, a team in a doubles tennis match. His dad holding him up in the air so he could serve the ball. He looked at the smile on his father's face, how broad and bright and full of life it appeared, even in fuzzy black and white Polaroid.

Boy, have I fallen into it Dad. I never thought it possible to be in such a mess. I wish you were here. But I guess that's my fault, now isn't it? All these years…fourteen or so I guess, and not a day goes by…

The phone rang.

Bennett jumped from his despondent funk and picked up. He expected to hear Hoosier's voice. Instead he got an official sounding voice asking if he was Bennett Ackerman. Bennett was hesitant to answer the question, wondering if it might be law enforcement.

"Hello…Is this the Bennett Ackerman residence?" the voice repeated.

"What do you want," Bennett said dryly.

"It's important I talk to Bennett Ackerman right away."

"Are you trying to sell something?"

"It's about his grandfather."

Bennett immediately felt a wave of shock bolt through his body. He sensed something was not right, even before the phone ever rang. "This is Bennett Ackerman, is he okay?"

"Bennett, this is Dr. Smithson with the Barnstable County Hospital…I'm afraid your grandfather has suffered a heart attack."

"Is he okay?"

"He's stable, but you should get up here as quickly as you can."

Chapter XIII

The sun's brilliant glare on the windshield exposed every imperfection the glass had ever absorbed. Why Bennett took notice now, and had not at one of the hundreds of other opportunities when he drove with a sunset blazing behind him, he didn't know. The trees, buildings, people, everything that passed by as he drove, appeared as though it had been dipped in a pool of intense light. A strong wind still gusted through the streets of New Bedford, tossing paper objects wildly into shrubs on the sidewalks and the entrances of storefronts. Life at the moment seemed different from what it had been before.

He turned on Interstate 195. Somehow he had managed to make two phone calls before rushing out the door. He called Hoosier first, getting only her answering machine. He left a message briefly explaining what had happened and that he would call her later in the evening. He next called DeWald. DeWald sounded very sympathetic and said it was okay for Bennett to go; he would take care of informing the local authorities. He told Bennett to report to him when he arrived in Barnstable and that he would have the authorities there keep tabs on him. He also told Bennett that despite this latest situation, that he, Bennett, must continue his pursuit of his own exoneration, as difficult as he may find that to be under the current circumstances, because time was no longer a variable, the DA had defined it for him. He had one week and one week only.

In some respects a deadline actually made Bennett feel better, after all, now he didn't have to worry about someone snatching him from his house when he was seconds away from discovering a piece of crucial information. The unknown had kept him up at nights wondering when— when would they come for him. In this sense he felt better. In other respects, he was terrified. A week?—and if he had nothing by then? Well, that would settle everything. He would be off to trial. The potential fifty or so years that lay before him, the ones he never thought about, the ones he assumed he'd be free to spend as he chose, had now been compressed to one week.

One week. Seven days.

Time was relative. As a child, one week before Christmas was an eternity, now, as he mulled it over, it was a flash, barely enough time to think. But think he had to, above the thoughts of losing a loved one, ignoring the pleas of cherished memories that pulled his thoughts to his granddad. He would guard his thoughts with blinders, what he chose to not think about, could not detract him.

He hit the brakes and pulled the car to the shoulder.

A tear welled in his eye as the car came to a halt. I can't do it Granddad. I can't ignore you. It may mean my week passes with me at your bedside, but that's the way it has to be. The pressure he felt inside, building over the course of a week, erupted and he fought back tears. It took minutes for him to recover from the waves of emotion that had swept through him. He wiped his eyes on his shirtsleeve and took a series of deep breaths. He felt better. Perhaps the tears in some metaphoric, if not chemical way, expelled the buildup of stress in his body. He shifted into gear and pulled back onto 195.

Life is tough, man. Life is tough.

The remainder of the drive to Barnstable was an interminable one. The day had turned to night. He found the hospital with less effort than he anticipated. He'd been there once before, the day he slipped on the rock, cutting his leg, and Patrick had carried him back to the house. Granddad had piled he and Patrick in the car and brought them to the hospital. After all those years, he recalled how they had got there. This time there was no minor cut to his leg—there was a major tear at his heart.

Brightness, a combination of powerful fluorescent lighting and every visible object adorned in the color white, flooded Bennett's eyes as he approached the front desk at the hospital. A chipper candy striper asked if she could help. Her ebullient expression remained constant when Bennett mentioned whom he was and why he was there. He interpreted that as good news. She directed him to a waiting area, where he picked up a magazine and pretended to read it. A woman with a child, heavy in fever, the child's hair wet in sweat, sat across from him, reading some colorful periodical with a perky half naked woman on the front. The child stared endlessly at Bennett.

A nurse poking her head out from behind a door called the woman and child back. Sometime later Bennett saw them leave, the child waving slowly at him as the mother carried it out the front doors. He sat alone, only the occasional passage of the transient staff as an indicator that time was actually moving. One doctor—looking individual after another came and went, each time Bennett raising his head expectantly, and lowering it dejectedly.

Bennett rose to stretch and contemplate whether to raise a stink, when a man wearing thick black rimmed glasses with brushy eyebrows sprouting above them, took a seat not far from Bennett and asked him to take a seat too. The man introduced himself as Dr. Smithson; Bennett shaking what felt like a clammy hand.

"His prognosis is good, I think, Bennett," the doctor said loudly and shook his head at length. "He's a strong man. We got him in just in time."

Bennett asked questions one would expect him to ask in the given situation. And the doctor gave him answers.

"He could die, I'll be frank with you. He could die tonight, but then, he could live ten more years—who knows? Right? None of us knows. That's in somebody's hands greater than mine."

Bennett asked if he could speak to his granddad.

"That depends," the doctor said shrugging, "do you mind talking to someone who can only communicate by squeezing your hand? Some people would just as soon not see a family member at all than to see them in that state. You know what I mean? Right? It's in hands greater than mine—what can I say?"

The doctor explained the zombie state as temporary, more of a result of the medication and trauma, than some permanent damage to him as a whole. It might be possible for his granddad to speak to him tomorrow, if all goes well. In the end, Bennett decided to peer in on his granddad from the door of his room. A tear escaped Bennett's eye as he observed his granddad sprawled out on a sterile, metal-barred, bed, with cold mechanical cogs and flywheels beneath and above, beeping, blinking electronic monitors hunched around his head like a group of curious robots. He crept in and pulled the thin blanket over his granddad's exposed toes. The doctor assured him he was very stable and would be watched closely. Bennett thought it over, knowing he would need to contact Patrick as soon as possible, and decided he would make a quick stop at granddad's house to get the phone number and call him and then call Hoosier. He left the hospital leaving his granddad's number as a place to reach him should they require to contact him.

He pulled into granddad's house, finding many of the lights inside still on. Of course, when having a heart attack, turning off the lights is probably the last thing on one's list. He stepped from the car and began fondling through his keys to find the house key when he heard someone approaching from behind.

"Hello, Bennett..." Bennett turned. "Don't act so glad to see me."

Bennett shot a nasty look in the man's direction. He saw enough of the man's features under the streetlight at the end of the driveway to make him out. It was the last person he wanted to see, or thought he'd see, in his granddad's driveway. He stood in silence as the man approached him—the man stopping directly in front of him. Inches from Bennett's feet, wearing a bristly face shadowed below a clump of tossed hair, were those feet attached to the arched frame of Detective Laykin. Laykin grinned, his foul breath carried in the mist puffed between his lips, uncovered in the heavy fog settling in. The two stood in silence, neither flinching, the harbor foghorn sounding periodically in the distance.

Bennett stepped back. "To what do I owe the honor?"

Laykin's grin dropped, followed by an outburst of laughter. When his laughter was under control he sniffed a few times and said, "Tell me Bennett, are you a fan of the finer things in life, you know art and so on?" A tick erupted in his brow.

"Just where are you going with this?"

"Well here..." Laykin moved over by the hood of Bennett's car. "Let me give you an example." He proceeded to draw a circle in the moisture collected on the hood of the car. He drew a larger circle around the first circle he had drawn. "Now tell me Bennett, is that a work of art or what?"

Bennett shrugged vacantly. The foghorn sounded.

He continued. "Oh I get it, you're not a fan of abstract art. Perhaps an explanation?" He didn't wait for an answer. "You see Bennett, this particular work of art is quite common in our prison system. In fact this particular picture is created just about everyday." He smiled, leeringly. "Haven't guessed what it is yet, have you?"

"This is ridiculous—"

"You see, my friend, this inner circle, well, it represents the size of your bunghole when you get into prison. And the outer one?—Well, the outer one, I'm afraid it represents the size of your bunghole after a day or so into your prison life. And you know what this all means? It means more crap will pass through as you roast in the electric chair."

Bennett turned his head away and shook it, pursing his lips. "You haven't answered my question, why are you here?"

Laykin's smile dropped and anger rose through his features. "You think I'm just trying to play with your mind?"

Bennett sniffed, still looking away from Laykin. "Why are you here?"

Laykin pushed himself away from the hood and walked around Bennett until he stood face to face with him. "I'm here, boy, to keep an eye on you! You aren't gonna kill any more innocent women, if I have to kill you to make sure it doesn't happen."

"Aren't you a detective and aren't you out of your jurisdiction?"

"I'm not on duty my friend, this is on my own time."

"Does Agent DeWald know you're here?"

Laykin burst out laughing. "Of course, your buddy, your best pal in the whole wide world!" Laykin clenched his teeth. "DeWald is nothing! He's freaking support personnel. If you're thinking he's your superman who's gonna come flying to your rescue, well, you better think again."

Bennett found himself clenching his teeth. A wave of adrenaline, flooding his muscles with surging energy, swept through his body. He felt himself drawn to the idea of purging the mounting energy though his fists as they recklessly hammered into Laykin's soft body tissues. He so wanted to relieve himself, to batter the man to unconsciousness, issuing bone-crunching pain while taking him there. He could see Laykin's eyes detect his locked jaw and clenched fists and smile knowingly. Bennett realized it was all a part of his plan. He wanted Bennett to strike him, to assault him in some fashion, providing him the justification to haul him in. Bennett swallowed hard.

Laykin swished his mouth. "You want to strike me don't you?"

Bennett nodded. "I'd like to bust your chops."

"Why don't you? I'm off duty. I'm wearing no badge, just you and me, man to man...Afraid I'll kick the crap out of you?"

Or maybe he thinks he can beat me badly enough to put me in the hospital and off the streets—

"Come on Bennett, just you and I..."

Bennett held his animal instincts in check for a moment and thought over the situation and how it was ridiculous. He sighed, saying, "I don't have time for this, I've got a grandfather on his deathbed," and turned to walk towards the house.

"Yep, I certainly pegged you right, you won't fight a man."

Bennett kept walking.

Laykin raised his voice. "Did I tell you—I questioned your grandfather earlier today. I wonder if that had anything to do...no, probably not."

This time the adrenaline jolted through his body at out of control speeds. Before giving it a thought, his instincts reacted, taking to large steps, jumping to the hood of his car, lunging headfirst into Laykin, who failed to react, like a deer caught in the headlights of an oncoming car. Bolts of energy shot down his arms and through his fists as rapid-fire punches pounded Laykin's face like a lump of bread dough. The release felt good, like scratching the itch that has relentlessly plagued one's skin. As his fists pounded like pistons in a two-cycle engine, he watched as the pallid skin at the other end began to smatter in crimson.

And then he felt a blow to his jaw, strong enough to send him flying back on his rear. He worked to steady himself and prepare for an attack from his enemy when he caught site of somebody else who wrapped Laykin in a bear hug and as Bennett rose to his feet, he felt somebody do the same to him.

Jutting his jaw in the mirror, he took notice of a pink patch on his skin and a dark red band inside it, Bennett shook his head, realizing the band was from Laykin's ring. It turned out the officers DeWald had arranged to watch him while in Barnstable proved to be of some value. It was they who broke up the fight. Laykin had explained to them, spitting blood that he too was a police officer and that he and Bennett were holding a discussion that got out of hand. Under the streetlight, Bennett could see the battering Laykin's face took, and at that point wished he hadn't beat the man quite so much. A few select punches would have been adequate. He rolled his hands before his eyes and he watched his fingers tremble.

He felt like taking a drink, but knew he had work to do, so he dismissed the urge and opted for a glass of cold water instead. It was only when he returned to the living room that he observed some things: a coffee mug laying on its side on the carpet, a brown stain projecting across the carpet from the mug's rim, a newspaper sprawled around the floor, and a pair of eye glasses, broken, lying next to the mug. He surmised what Granddad had been doing when the attack struck. He picked things up as best he could and returned, opening the top drawer in his granddad's desk. He found inside, an old-fashioned address book, made of steel, with a glide bar on top that slid within the alphabet painted to either side. He slid the bar to the letter 'A' and clicked a button that swung open the face, revealing the 'A' page. He scanned for and found Patrick's name, but realized right away the address was tremendously out of date. As he maneuvered the thing to put it back in the drawer, it sprung open on him and out jettisoned a yellowed newspaper clipping. He unfolded the delicate newsprint and read.

A half-dozen homes in the Wilburs Point area off of Sconticut Neck Road in Fairhaven suffered property damage late last night as the result of a destructive spree lead by juveniles. It is believed the juveniles earlier in the evening attended a Halloween party at the same location. It appears the annual party got out of hand as some partygoers terrorized the area,

*breaking-out car windows, spraying graffiti on the sides of homes, and
destroying gardens. The apparent leader of one group is reported to have
explained the incident by stating, "We just wanted to have some fun—what's
the big deal?" Detectives believe the boys were under the influence of
drugs. "The self-proclaimed leader of the group thought he was a pirate," a
detective was overheard saying...*

The article was dated the same day he was found injured by the side
of the road on Sconticut Neck. Because the article was from a local New
Bedford paper, he wondered how it had made its journey to Granddad's
desk drawer. Someone had been its courier. The rogue groups caught his
attention. Perhaps he had fallen victim to one of the marauding gangs, at
least whoever cutout the article may have thought so. And the leader—a
pirate...*a pirate?* It had a familiar ring; at some subconscious level he
sensed it. Though he lacked a rational reason for discerning this, he knew in
the dark corners of the sensation lurked something more than fantasy.

It occurred to him mid-thought that he had not made a single phone
call, and after all, Granddad was ill. He returned to the desk drawer and
combed through its contents, soon discovering a much more recent address
book. He found Patrick's number and dialed him. After five rings he got an
answering machine. He left a detailed message, hung up, and dialed
Hoosier. He could sense the concern in Hoosier's voice as she questioned
him about the state of his granddad. She asked if she should drive up, but
Bennett thought she could be of more help back in New Bedford. He told
her about the one-week deadline and heard nothing but white noise for a few
seconds before she replied optimistically that they could get the answers
they needed in a week.

Bennett hesitated. "I don't know Hoosier, with Granddad in the
shape he is, he needs me. I decided on the way up that come hell or high
water, I'm sticking by his side until he's back to his old self."

"You said yourself he's stable."

"For now he is."

"Look, I know how important your granddad is to you, but this is
your life we're talking about. I don't mean to sound cold, but you need to
take care of yourself. Patrick can stay with him."

"I can't do that, Hoosier."

The line carried nothing more than a low white noise and jumbled
cross talk—some woman talking to a man named Harold about her
overweight cat—for the few moments until Hoosier spoke. "This wouldn't
have anything to do with your father would it?"

"What are you talking about?"

"I'm talking about guilt."

"If you're asking about abandoning a loved one in the past, then the answers yes. I did abandon him Hoosier. I just laid there, like some stupid kid—"

"You didn't know he was there."

"I should have gotten on top of the boat, he would have seen me immediately, but I was too scared."

"You were a kid, and now you're attempting to judge your actions as a child with your adult mind—that's not the way things work."

"I'm staying with Granddad, Hoosier."

"You're making a big mistake and I hope you realize that before it's too late."

"That might be, but it's my mistake to make."

He could sense Hoosier's frustration in her voice. He wouldn't blame her for telling him to go to hell.

He heard her sigh. "Okay...we'll just have to do the best we can," she said.

"So you're still willing to stay in the game?"

Hoosier paused. "I'm in love with you, Bennett Ackerman."

The words stunned Bennett. Not that he disliked hearing them, but he felt emotionally saturated. The fear he felt for his granddad and himself, the anger for Detective Laykin, and now the words from Hoosier. It was too much to occupy his mind at one point in time. A strange dizziness, spun by a whirlwind of emotion, overcame him, rendering him speechless.

"Bennett?"

He tried to recover—to think—to come up with some sort of response that made sense. Mentally, he had earlier acknowledged a strong affection for Hoosier, but emotionally, for the time being, the feelings were lost in a nebulous cloud of sensations in turmoil. His hands still shook.

"Did I frighten you? I guess I shouldn't have said what I did, at least not now."

He decided to respond with what he knew to be true and not what he felt at the moment. "No, that's fine. But...well I think...I mean I know—I feel the same way about you."

"That's why I'm staying in the game, Bennett."

He told Hoosier about his encounter with Laykin. It appeared to shake her up; nervousness crept into her voice when asking questions concerning the fight.

Then she informed Bennett that Tanya was a cop.

A cop? She suspected me all along...that's why she was with me...she thinks I'm guilty...was I taken for a ride! What a fool...I am such a fool!

He questioned Hoosier thoroughly over her detective work. Could the IDs have been forged? But why? The more the idea was pondered, the more it fit. It explained why an attractive woman would have found interest in him and why she felt it necessary to become familiar with his friends. It explained their conversations and her questions—never direct to make them obvious, more like side comments that led into questions, the type of behavior one would expect from someone undercover probing into areas first dulled with an antiseptic of charm in an attempt to go undetected. It didn't explain what he found on her computer, unless what she had was evidence taken from some other computer. What else did she have that might be of usefulness? DeWald had to know about her and had not said a word, nothing—he wasn't sure whether that particular information upset him or not. Obviously if DeWald told him, he risked blowing her cover and in the end that could hurt the investigation. And, maybe, DeWald wasn't entirely sure he was innocent, in which case, he could not possibly give Tanya away. The whirlwinds spun with even greater velocity.

"Does she know you know?" Bennett asked. What a fool…how could I have been so stupid?

"I don't think so, but if she ever goes up into her attic, she will know. I left it a mess."

"I think we'll have to assume she'll know."

"What do we do?"

"I'm not sure. Let it be for now I guess."

"What should I be working on?"

Bennett thought for a moment. "Let me think on it." I am such a fool…

When Bennett got off the phone with Hoosier, he immediately called the hospital and asked to speak with Dr. Smithson. A nurse informed him that the doctor was not available, but assured him that his grandfather was doing fine and that the doctor would call him when he had a moment. Based on the information, and Hoosier's admonishment still ringing in his ears, he made the decision to not return to the hospital right away, but to take a moment and unload his laptop from the car and connect to Pequod to check his e-mail. He hadn't for a while and it would only take a matter of minutes.

Minutes later the modem in his laptop squealed as it negotiated a connection with Pequod. A dulcet tone informed him he had mail. When opening his web email for his covert account, a series of 25 letters stacked into his inbox. His new website had generated interest. He started with the oldest message, read a few lines, was overcome by disgust, and trashed it.

The next message fell to the same fate—and the next, the same too. The fourth caught his interest.

I am he who would fill your slot. I am Ahab. I am the exception to my comrades, whom you have chosen to occupy your page, for unlike them, I kill not of my own desire, but at the behest of the great beast, whose desire it is to devour the bodies of the wretched—whom are inclined to force their bodies to endure malicious chemicals—and expel them in the rapture of a fine perfume; they are reborn the ambergris. The great whale swimming the seas of cyberspace hungers for these heathen souls. It is for his hunger that I kill. Lest you succumb to that abominable assertion that the great Mocha Dick, his aged body haggard and weak, died at the hands of a Swedish whaling crew, of which the harpooner, exerted a lazy toss of his harpoon to subdue the creeping beast, I am here forthwith to explain it as all a great deceit, manufactured by man to deny this whale his rightful claim to deity. He lives! And it is I who opened the portal, permitting the beast, with one great thrust of his immortal flukes, to jettison from the ocean depths of the South Seas to the electronic currents of Bell South. And it is there that he now roams. And it is I who feed him. Alas, when I first undertook this duty, I feed him electronically with images, but henceforth this past March, he refused them. And so it did occurr to me that I should feed him what is real, for though an image captures a view of a reprobate, and is encased by the ambergris, the transformation is without efficacy. Real renewal requires real bodies. Finding a likely female, for the ambergris requires that feminine touch, I marked her forehead to indicate she belonged to him, and I so fed him the woman vicariously through the mouth of a humpback sow. And likewise I feed him today. And once nature takes its course and brings the morsel to his bowels, he sings a sweet song, the song that expresses his great pleasure at having the ambergris dwell within. He carries them all in his bowels, and as of late, was seen swimming not far offshore of that island occupied by the Typee. But I beseech thee not to engage in holy pursuit of this deity, for his wrath is great and death lingers only inches from his flukes. Heed my warning.

Bennett sat spellbound by the words. Was it a hoax letter? He pondered the thought momentarily, but realized the source of the message had to be genuine. Nowhere at his site did he mention anything about the New Bedford killer having any connection to whales and DeWald and company had successfully hid the information from the public. Only the killer would know about the connection. Only the killer knew about the tattoos on the forehead, as he had seen in the digital photo. Shaking his head, Bennett read the message again. The letter confirmed his theory, the one DeWald would not deny.

For the first time he had a window into the man's mind—some clue as to who he was and why he did what he did. What he found sent tingling down his spine, a shiver spawned by the cold empty darkness that he perceived as dwelling in the man's mind. It was a conscious state that Bennett understood was there, a primordial set of synaptic energy, rarely fired, buried deep under civil suppression, and as he pummeled Laykin he sensed himself lifting the suppression, calling out to what lay below, and tumbling to that same cold empty darkness. He had awoken from dreams and seizures, drenched in sweat, conscious of the fact that his mind had dared to journey there.

He had few doubts the man was sincere, and fewer yet that he was lost in a world known only to himself, some mystical world he had created, perhaps with the aid of technology, to escape whatever was chasing him. Tragically, it was a world that contained passages to the real world, passages through which he could grip his victims, pulling them screaming into the crypt of his tortured mind—a human wolf spider.

Bennett shivered again, knowing that the answers he needed to find were in this man's world and that he would, of his own volition, enter the passages to get there. He took a moment to add Typee to Dory's search list. He copied the message to two floppy disks, remembering DeWald's words of wisdom—document everything, information being the key to his freedom. The next ten messages were discarded as the first three. The one following caused his jaw to drop.

Don't particularly like your site, but came upon it when searching for information to write a report required for my History class on the Civil War (another site overloaded with keywords that have absolutely nothing to do with the site's content, a huge pet peeve of mine. Anyway, your image of the tattoo caught my eye. At first I was curious as to how you acquired the image, because the man I knew who had it, wouldn't want anything to do with your site. But then I thought it's probably just a coincidence because the man I knew is dead. He was a shop teacher at my high school. He was weird, but I think most everybody liked Paul Kitt.

Chiclets

Paul Kitt—Patrick's long time friend! And he had died recently according to Patrick. It made sense. It also made Paul out to be the whale killer. The last one had gotten the best of him. His arm must have become entangled in the line and then one of two things happen: either his body was somehow secured to the boat and the whale ripped his limb from his body leaving him bleeding to death on the deck of the boat, or, and more likely, the whale pulled him overboard and dove with him in tow to the depths, perhaps swimming through the equivalent of a picket fence, and dispensing with that part of Paul's body that would not pass through the rocks.

Pinning the whale killings on Paul was incompatible with his theory that the killer of the whales and the killer of women were one and the same individual. The email from the individual calling himself Ahab, the assumed killer, was dated after Paul's death, and even if for some miraculous reason the email had been sent earlier and delayed, Paul could not possibly have written it, the words, having more than four letters, were not the types of words he would have chosen. He had to conclude Paul worked with the killer, probably as some stooge. That or the killings of the whales out at sea were not related to the killings of the women on the streets of New Bedford. Bennett's thoughts moved on to Patrick. How could he not know? He had said he was with Paul when he died, but that had to be a lie, unless…

Bennett awoke with a startle. Sitting in a chair, his head snapping up out of some reflex reaction, and a bustling alien environment around him. He rubbed his eyes and brought himself to the realization that he was in his Granddad's hospital room. He could hear the equipment performing its duties in the background. He groggily rose from the chair and walked over to his Granddad's bed. As he stood over him, patting his hand, Granddad opened his eyes. Bennett felt his spirits soar.

"Granddad?" he said softly.

Granddad's eyes blinked and his mouth opened. His lips moved but what words he was attempting to construct remained in his mouth. Perhaps, the muscles he required to speak were still under the control of his medication. Or maybe his vocal chords were slow in warming up.

Bennett bent over and whispered close to his ear. "Can you hear me?"

He nodded and attempted a smile, but fell short.

"You don't have to talk, just nod."

He spoke anyway, his voice hoarse. "Guess I had the big one, huh?"

"You'll be okay. This is a minor inconvenience for the man who used to carry me on his shoulders to the candy store."

He managed a real smile. "Shouldn't…you be…at work?"

"Granddad, you've had a heart attack."

"Bad heartburn."

"Nope, it was the real thing, a genuine, grade A, 100 percent heart attack. How are you feeling?"

Granddad rolled his eyes around the room. He cleared his throat, which appeared to allow him to speak with greater ease. "Pretty good…despite all these gadgets…they've got me hooked into." He coughed as best he could. "Does Patrick know about this?"

Bennett nodded and explained that he had left a message. "I'm sure he's on his way up now."

He cleared his throat again. "You're the best grandson a grandfather could ask for, do you know that?"

Bennett smiled. Here Granddad was cheering him up; it should be the other way around.

"Yes indeed," his granddad continued, "ever since you were a boy I knew there was something special about you. You have a special talent for what you do and have had it since you were a child, though you are not aware of it. I could see the way you took things apart and reassembled them. I could see the genius in you."

"Well obviously I grew out of it."

"That's not true and one day you'll know it's not true."

"Well, I appreciate your confidence in me, but I think its time to stop talking and get some rest."

"And you had such an affection for the sea," Granddad continued, ignoring Bennett. "You and Patrick, always down at the beach, doing something that involved the sea. That's why I called you sea boy—or at least that was the English translation. I remembered very few words of your grandmother's native tongue, but I remembered two that fit you. I wanted to give you something from her, so what better than a name in her language. Let's see, how did I say that in Russian—"

"Well Granddad, you seem to be looking much better this morning," Dr. Smithson said, standing at the base of the bed reading his chart. "Looks like you spent the night here, Bennett."

"I look that bad, huh?"

"You ought to go back to your grandfather's house and catch a few winks. As you can see the patient is doing great."

Bennett nodded. "I want to spend a few more minutes with him. Hopefully my cousin will show up soon to take over."

The doctor checked a few things and asked Granddad a few questions and then left the room.

Granddad gripped Bennett's hand. "I'm telling you these things because I may not get another opportunity."

Bennett shook his head. "Of course you will."

"Maybe, maybe not, but I want to say them while I still know I can. Now what was that name in Russian—see how sharp my mind is, I can remember what I was trying to remember, if that makes sense."

Bennett grinned. "As sharp as a tack…"

"Morye…Morye…oh yes, that's it…Morye Malchik."

Bennett cocked his head. Why did that name sound familiar? "You called me that?"

"Up until you got into high school, then you begged me to no longer call you that. I think Patrick kidded you no end about it. He twisted the words into something vulgar."

"Did I ever use the name?"

Granddad rolled his eyes in thought. "I think you did. Before you became self-conscious about it, I think you used it."

Bennett repeated the name in his mind. Morye, Morye, Morye…Yes! It was the name of the author of the code he unraveled for the whale files.

But it made no sense. Either it was a bizarre coincidence or whoever wrote the code knew him when he was young and knew him well. But who could that be? A sick sensation developed in his stomach. Patrick knew Paul and he. A human bridge to cover the gap between the whale killings and the…It couldn't be though, Patrick could not have written that email—it didn't fit his personality. But did he really know Patrick well enough to draw that conclusion? Could it be? But why? Out of revenge?—Patrick thinking it was Bennett's fault that his uncle had died? Bennett walked back to his chair and slumped into it.

"Are you okay, son?" Granddad asked, turning his head to look over at him.

Bennett nodded.

"You look like you've seen a ghost. Don't tell me I look that bad."

"Maybe something worse," he muttered. Without conscious thought, he slid his hand into his front pocket where it came in contact with a thin object. He felt around it when it occurred to him that he had folded up the newspaper article he discovered the night before and placed it there. He looked down at his Granddad and observed the man's eyes, beneath his white caterpillar brows, concentrating on his face.

"You get in a fight, Bennett?"

Bennett averted his eyes. "Fight? No, but did take a slip in your bathroom."

Granddad sighed. "Sorry to hear it. Thought I had the tile covered better than that."

"Do you recall ever seeing a newspaper article about an incident at Wilburs Point that occurred the night before my accident?"

Granddad's eyes finally left Bennett's face. He focused on the ceiling. "Let me see…" He pursed his lips. "Yes, as a matter of fact, I do recall seeing something. It was left behind after one of your visits; you had some friends with you. But why didn't I mention it to you? I guess maybe I thought it was yours. Where did you find it?"

"Your old mechanical address book."

"Ah, you know what? I'll bet I stuck it in there to remind me to tell you about it next time I called. I probably put it under the wrong letter of the alphabet."

"Who was with me on that visit?"

Granddad gradually moved his head from side to side. "That, Bennett, I'm afraid, has escaped the old memory banks."

He listened distantly as two nurses took turns sharing stories about their patients, their long hours, and the head nurse who they both hated. It was late morning, but Bennett still felt compelled to drink a strong cup of coffee. He sat on a cement bench in an open area between two of the hospital's wings, gulping the hot java. He desired the fresh air to think, but soon realized, as both nurses lit up Virginia Slims, that he had chosen a designated smoking area for hospital employees. He had hoped the caffeine might intensify his ability to make sense of the great amount of key information that had wandered his way as of late. Though he had voraciously sought out the information, what was returned to him was more than he had ever expected. Now he felt compelled to tie it all together— wrap it up into a nice neat package and drop it on Laykin's head.

He considered the email from the killer. The man, at least he assumed it was a man, knew his whaling history, had obviously read Melville, or knew of his writings, and was very conversant with computers. The whaling history and Melville did not rule out too many folks, after all New Bedford was steeped in it. The computer expertise, that narrowed it somewhat, but Internet aware individuals were becoming a majority. Granted the knowledge the killer had exceeded that of most, but was nothing a studious kid in his basement with his PC, a set of RFCs, a compiler, and soda and pizza, couldn't accomplish over time. Kind of like the monkeys in a room with a typewriter, but more direct. There would be no particular reason for anybody in New Bedford to be aware the kid was there, but the hacker sites across the globe would, and he'd probably be a hero in those neighborhoods. But *Dory* was searching those sites, and had returned empty-handed.

He was not getting very far with his profiling exercise, so he decided to consider Mocha Dick. What first occurred to him was that maybe Mocha Dick was really nothing more than a figment of the author's imagination, an envisioned higher force that absolved the author from responsibility for his hideous crimes—some illusionary beast that roamed

271

his mind and not the wires and cables of the Internet, as he professed in his letter. But it could be real; it could be a part of the set. After all, the man had produced whale and harpoon files along with seagull processes. One could deduce that Mocha Dick might exist as a process—sort of a king reigning over all the other players. As a process, he could potentially navigate the Internet, but to move from location to location required there be something waiting at the next destination to which he could send himself. The waiting piece could be a player in the set, perhaps the seagulls, or the malicious manipulation of a benign process. But beyond this, there were such things as firewalls. If Mocha Dick were some super process, he could not just saunter freely from site to site; firewall technology would ensure that. But if he were aware of a hole in a process on the inside of the firewall, and that the firewall permitted traffic to that process, then he could swim on through. If he were to discover Mocha Dick to be a super process and Mocha Dick did indeed contain within himself information of the killings, as the author had suggested, then his pursuit and detection would be very worthwhile—but how to capture him, especially if he were in some safe harbor behind a firewall—that was the next logical question. Perhaps the answer lay in luring him to a place where he could be readily trapped—

Something caught Bennett's eye as he peered through the tall glass of the wing alongside him. He observed a wiry man running towards him, the man's long hair bouncing behind his head. Bennett recognized the man as Patrick. He shot up and slid through the glass door behind him, stepping right into Patrick's path.

"Bennett," Patrick said, breathing heavily. "How is he?"

Bennett put his hands up. "He's fine."

The worry on Patrick's face remained. "His room is just down the hall, right?"

"Yes."

"Alright, I'll see you—"

"Patrick, I desperately need to talk to you—"

"Give me a break, Bennett! Granddad's on his deathbed!"

"Look he's okay. When you get done, just meet me outside, okay?"

"I might," Patrick said as he sprung past Bennett.

It seemed like hours had passed, but when Bennett looked at his watch, it had been only 45 minutes. He had grown tired of analyzing the information; the missing pieces he seemed to encounter, and the terrifying conclusions he found himself forced to draw, battled against him. He was on the verge of taking himself up on the thought of a drive when he saw Patrick approaching from the wing. He met Patrick inside and suggested they take a walk outside the hospital. They followed a sidewalk that cut diagonally across the grounds to a far street.

"Wouldn't you agree he looks pretty good," Bennett said, patiently.

Patrick pulled sunglasses from his pants pocket and slid them on. "I just want him out of here."

"He will be soon enough."

The two walked in silence. Bennett took the opportunity to broach the subject he so desperately needed to discuss with his cousin. He had thoughtfully rehearsed what it was he wanted to say, but had never arrived at a comfortable delivery. In lieu of that, he decided it best to just jump into it.

"I'm going to tell you some things," Bennett began, "that you're not to speak to anybody else about…agreed?"

Patrick shrugged.

"I'm going to be perfectly candid and I expect you to be the same."

"So what's this about, Bennett? If it's some confession, I'm not Dear freakin' Abby."

"Always a wise crack," Bennett muttered. "It's about you and me and the trouble we're in."

Patrick furrowed his brow. "We're? Don't drag me into whatever this is."

"Let me first say I know Paul was killing whales." Bennett was amazed to observe no reaction out of his cousin. "Did you help him?"

Patrick stopped. "Look, I thought you were going to be telling me some things—not asking."

"Don't evade the question."

The two stood almost halfway along the diagonal sidewalk. Nobody was around for hundreds of yards.

"Let's cut the crap—they think you're killing the whales, right?"

"What would make you think that?"

Bennett observed Patrick noticeably thinking. "Why else would you be trying to pin it on me?"

Bennett went fishing. "Pinning it on you? It wasn't me who left someone else's boot behind—with a receipt in it I might add."

Patrick pursed his lips. "So who made you freakin' Sherlock Holmes?"

Bennett lowered his head. "It goes further, Patrick. Further than you planned. Look, I know you'll never forgive me for what happened to dad, and you want to hurt me, and pinning the whale killings on me would certainly accomplish the goal, but I also know you don't hate me so badly that you would want me dead."

"What are you talking about?"

273

Bennett looked around him. He could see the local PD car in the lot, the two individuals still inside. They were out of hearing range. "The police are of the opinion that whoever is killing the whales, is also assisting the serial killer, or is the serial killer himself. Do you see where that leaves me?"

Patrick rubbed his forehead. "They think they're related..." He gritted his teeth together. "Why?"

"They're not sure, but I think I know. I've been communicating with some guy on the Internet that goes by the name of Ahab. He's warped. I'm sure he's the killer and I think he's been burying his victims in the whales. Purifying them, as he terms it, into ambergris."

Patrick looked away. "Whew..." he whistled through dry lips, and swallowed.

"They've concluded it's me. I've got one week. One week without proof of my innocence, and I'm going to trial for the murders." Bennett could see himself in Patrick's sunglasses, but could not see Patrick's eyes behind the glasses. "Did you help Paul?"

Patrick slid his hands in his pants pockets. "I'm clueless—"

"For crissakes Patrick, you can save my life here! This is no time to be hiding behind lies!"

"Paul was a freak, man. He lived with his mother. I wouldn't put anything past him."

"As close as you and Paul were, how could you not know?"

"You talk like I held it for him when he drained his lizard. The man had his own life, and spent his own time, not always with me. How do you know he's not the killer? Let's pin it on him. He's dead, he won't care."

"It's not that simple. And besides, I know he's not the killer."

Patrick swallowed. "This sounds serious."

"The electric chair usually is."

"Yeah..." Patrick chewed on his tongue.

Bennett wiped both hands over his face. "Patrick, I beg of you, if you're involved in this tangled mess, if you've set me up, I don't hate you, but *please* work with me, together maybe we can get out of this thing with neither of us going to prison. If ever there was a time to reconcile, that time is now. I'm giving you one last chance to level with me. Are you involved?"

Patrick adjusted his glasses. "No." He gritted his teeth. "Focus on the serial killer, the whale killing business means nothing. If you've got a week, don't waste your time with the whale stuff."

"But they're linked."

"How do you really know that? So you catch whoever is killing the whales and they're not linked, big deal, where are you then? Are you going to plea bargain out of a capital crime because you caught the guy?" Patrick removed his glasses. "If you'll promise me you'll spend your time going after the killer and not waste your time with the whaling business, then I'll take good care of Granddad."

"Of course, but I can't leave him."

"Bennett, go."

"He needs me—"

"Go!"

Bennett found himself open to relenting and nodded. The inexorable tick of a clock chambered in his head. Perhaps everything would be okay for Granddad. The name Morye Malchik overtook his thinking.

"Answer just a few more questions."

Patrick sighed. "If I can…"

"Do you own a PC?"

"No."

"Do you recall what Granddad used to call me, you know, the Russian name?"

Patrick scrunched his face and smiled. "Something like moron malt—"

"Close, more like Morye Malchik. Anyway, there is another angle to this killer. He's written some sophisticated programs that he's enlisted to the cause. When I decompiled his code I found Morye Malchik as the author's name. Now you tell me, how many folks knew about that name?"

"Don't look at me. As I told you, I know nothing about computers."

"Who then?"

Patrick shrugged. "I don't know—there were a few neighbor kids who must have known. After all, it wasn't some great secret."

"Do you remember any of those kids?"

"Not really."

Bennett nodded thoughtfully.

"What about you?" Patrick said, pointing at Bennett.

"Huh?"

"You probably don't recall, but I do, how you used to sit in our room, a nice day outside, and write your freakin' game programs."

"What games?"

"How do I know? I wasn't a geek like you. I left you in there by yourself."

"Well, it can't—"

"A pirate game—"

"What?"

"You wrote some stupid pirate game. You played it with your friends online."

"I did?"

"You don't remember? You spent hours hunched over that PC of yours. I don't know whom you played with, but it was open to anybody. I think you even advertised it."

Bennett looked stunned. "I don't recall any of it."

"You stored the stuff on floppy disks. I remember you having a heart attack when I accidentally laid a magnet next to your precious disks."

"So what are you saying?"

"Maybe it was you who actually wrote the stuff."

"That's crazy. And even so, how did he get it?"

"You're asking the wrong person."

Bennett went speechless.

"Now my turn for a question," Patrick continued. "Who did you tell the story about my killing a dolphin, to?"

"Huh? The dolphin story?—I don't know, I guess a few people. What does that have to do with anything?"

Patrick shrugged.

Bennett walked with Patrick back to the hospital. His cousin's suggestion that maybe he had authored the code, lingered in that same place in his mind where song titles resided that he could not recall for music he heard on the radio. He wondered if it could be, and that if so, how could he determine if it was the case? He knew the place to start. He would go there next.

The dogs playing pool in the picture failed to invoke memories for Bennett. He stood in the center of the room he and Patrick once shared at Granddad's house. In the room were twin beds, a desk and chair, rusty colored shag carpeting, pictures and photographs on the walls, and a nightstand with a lamp, in between the two beds. Bennett looked around the room in hopes that something would click, that some memory would spring forth, something he could use for footing that he might leverage to recall another memory. He sat at the desk, the one he supposedly spent hours at, programming his computer and playing games. He searched the drawers and found them all empty. He stared at the wall almost grimacing in effort to bring back a memory. Nothing happened. He went over to a closet and searched it, but in the end, found nothing more than some of his Granddad's winter clothes and shoes. He felt silly, but moved over to one of the twin beds and laid down. He wrapped his hands back behind his head and stared at the ceiling. A blue stain fell into his line of site.

Bob Hope.

Like the uniquely shaped cloud in the sky that one ponders from a beach towel, and decides it looks like something, the stain with its sloped proboscis, brought home the image of Bob Hope to Bennett. He remembered it, vaguely, but enough that his mind pulled onto a track. He could picture himself lying in bed at night unable to sleep and, with the aid of moonlight, staring at it, and using it for mental exercises, attempting to determine how he could move the stained square to the corner, by shifting other tiles, made possible by a missing tile in the opposite corner. His eyes combed the particle ceiling tiles, stopping in the far corner. There was something about the corner that intrigued him. Still unaware as to the draw, Bennett got up from the bed and walked over to the corner. The tile there was smaller than the rest, cut to fit, and had what appeared to be chips along one edge. He grabbed the chair from the desk and moved it to the corner. Standing on the chair he carefully pushed the tile up and moved it back to his right where he gently set it down on top of another tile. He kept his hand in the resulting hole and felt along the neighbor tiles. His fingers stumbled over what felt like a thick stack of papers; his touch skated across the top sheet. He pinched the stack between his thumb and index finger and dragged it to the opening. The papers were too wide to slide through the hole so he rolled them and then pulled them through. When he had brought them down to eye-level he allowed them to spring flat. A grin spread across his face as he eyed the partially naked woman posing on the cover of an old July Playboy. At least in one sense he was a normal kid. He shook his head and allowed the magazine to drop to the floor. He reached back up into the hole and felt his way around on the other tile. His fingers collided with what felt like a box. Knowing he could not grip the wide box, he used his fine engineering mind to push up one end of the tile it rested on, and causing it to slide down towards the hole.

A box of floppies dropped into his hand.

They were the old kind—the 5 and 1/4 inch type that were no longer used. He opened the box and pulled a floppy up from the rest. The label read: *Pirate Treasure*. He also found his name and a handwritten date of over fifteen years earlier. He began to wonder over what he might discover on the disks and paused—he would need to find a machine that read the old format before he could get any answers. He still had his old machine, he had kept it for posterity, but he could not remember ever using it and thus was not sure it even worked. The problem was the machine was at his house and Granddad was here.

He revisited the issue of where he belonged. Hoosier had told him to put himself first and Patrick had done the same. After all, Patrick was here and could call the minute something happened, and I pray it won't. He flipped through the floppies, all the while thinking about the answers they might contain, and then made up his mind. He would drive home.

Even after crossing the Bourne Bridge, the guilt Bennett felt for leaving his Granddad had not been relieved by his hour-long effort, made during the drive back, to convince himself it was an okay thing to do. What made things worse was he could not tell his Granddad the real reason for his sudden departure, substituting in a lame excuse that left him feeling that his granddad might interpret his sympathy as nothing more than a shallow gesture. In the end he decided to tell him there was a problem at Pequod that needed to be addressed right away. Not the kind of excuse one would use to leave an ill loved one, but he had nothing better.

It gave him great joy, when finally reaching his home, to see Hoosier's car out front and her stepping out of it to greet him with open arms. He called ahead to inform her he was returning. He had done the same with DeWald—something he decided he needed to make a habit of doing. He didn't inform DeWald of his encounter with Laykin, not fully sure why, but maybe because he decided that that was something just between himself and the man. Besides, DeWald couldn't miss Laykin's mottled face.

He and Hoosier embraced as he stepped from his car, she whispering in his ear he made the right decision to come home, and he whispering back that time would tell. Minutes later both were in the house and rummaging for his old PC in his guest bedroom, a room, which years earlier began evolving into storage area. After passing a series of empty boxes back to Hoosier, who then tossed them out in the hallway, he uncovered the box containing his old system. When he bent over to pick it up it felt heavy; at least something was inside. He carried the box to his living room where, after Hoosier cleared off some magazines, he set it down on the coffee table in front of his sofa. He painstakingly removed the packing covering the old equipment, aware that this was not the time for something to slip and crash to the floor. Minutes later, all devices were taken from the box and resting on the same coffee table. He asked Hoosier to cross her fingers as he connected each device and plugged them into a power strip. They both said a prayer, and he toggled the on/off switch. The machine made a high-pitched beep and began emitting a low hum. Gray colored text scrolled down the screen, suddenly stopping. Bennett rubbed his chin, contemplating why the boot-up process had halted.

"Is something wrong?" Hoosier asked, sitting close to his side.

"Boot-up sequence suddenly stopped. Now why would that happen?" he said, troubled.

"Maybe the hard drive is bad—"

The worry vanished. "Thank you, Hoosier!"

"Huh?" Hoosier said, cocking her head.

"This old machine doesn't have a hard drive, it must boot off a floppy, I totally forgot!"

Bennett quickly flipped through his box of disks until he found one marked bootup. He slid the floppy into the machine's drive and restarted it. This time the machine came all the way up.

Bennett rubbed his hands together. "Well Hoosier, here goes nothing."

He inserted the disk marked Pirate Treasure. After the machine acknowledged his disk, he listed the files on it.

```
anchor.c
banner.txt
crypto.c
deliver.c
images
intern.c
key.c
pigeon.c
pirate.c
pship.c
readme.txt
rules.txt
snavy.c
tchest.c
utils.c
```

Usually when delving into code, Bennett never felt it important to look at the standard readme file, however in this instance, where time was of the essence, he decided to oblige it by opening it first.

```
# Pirate Treasure Readme file
# Author: Morye Malchik (aka, Bennett Ackerman)
# Source file descriptions:
#__FileName___|_____Description_____
# tchest.c       Treasure chest. Can hold encrypted payload. In this case
                 pcx files with images of coins, gems, ambergris, etc.
# pship.c        Pirate ship. Super process that carries treasure chest and
                 leaves them hidden at user's choice of location.
# anchor.c       Anchor for pirate ship. Will allow pirate ship to return to
                 previous ports or export treasure to the port. One anchor
                 is left at each port visited.
```

# snavy.c	Spanish Navy ship. Enemy of pirate ship. Can sink pirate ship sending all its treasures down with it.
# pigeon.c	Courier pigeon. Sent out ahead of pirate ship to retrieve information on particular ports. If pigeon returns with information that port is okay, pirate ship can export treasure to the site, or send itself.
# pirate.c	Foot-in-the-door at remote site. Once in place, it listens for the arrival of email addressed to root and if it finds a key in the header, steals the email and executes the commands in the message body. A pigeon can be sent as an attachment and be launched by pirate. It is a comsat replacement.
# intern.c	Helper app that internalizes payloads into treasure chests.
# deliver.c	Helper app that sends treasure chests to remote ports, without aid of pirate ship.
# key.c	Unlocks treasure chest to reveal payload by decrypting payload in treasure chest. The result is unencrypted payload in a file.
# crypto.c	Encryption algorithms to encrypt payloads.
# utils.c	Common functions shared by other files.
# food.scr	File containing commands for pirate ship. When it comes to port for food, the commands in the food file are processed.

Hoosier read the file out loud, squinting to read the small text. "Not really much help, is it?" she concluded.

Bennett continued to read the screen. "On the contrary, it's of great help. I think I know what our friend did. Let me draw a mapping." He grabbed a pen and piece of paper and jotted down the following:

```
tchest   -> whale file
anchor   -> unknown
snavy    -> unknown
pigeon   -> seagull
intern   -> unknown
deliver  -> unknown
key      -> harpoon
crypto   -> support module
utils    -> support module
pship    -> mocha dick?
pirate   -> (comsat) email notification replacement
```

"Do you see it?" Bennett asked, smiling.

"I think. Are you suggesting he extended your code to build his sick little world?"

"Extended my code? I'm impressed."

"I took a few computer classes in college."

Bennett rubbed his eyes. "But, I don't think he even went that far. It appears to me that he used my code right out of the box and simply renamed it."

"My gosh, who'd have ever thought…"

Bennett continued looking at the file when suddenly he spun around to face Hoosier. "Of course!"

"What is it?"

"That would explain the weird messages in the system logs that both Griggs and Adam found. The treasure chest message came from my game. Apparently our friend missed some things when he renamed!"

"But how did he get a hold of your files?"

Bennett furrowed his brow. "Good question…" As he brought up the other files to read, he found his answer in the banner.txt file.

I have made my source code available on blackbeard for any of you out there who think you can improve the game. The only requirement is that you document the heck out of your code, following my standards, and that you let everyone know what you did in the upgrades.txt file located in the same directory. I believe this can only result in improvements to our game.

Bennett Ackerman

"He must have been one of the players," Hoosier commented.

Bennett flipped through the other disks in the box and stopped when he came to the one labeled Players. "I may soon have an answer for us," he said, gratifyingly. On the disk was a file called players.txt—he listed it:

jolly roger	- Bennett Ackerman
redbeard	- Mark Mitchell
blackbeard	- Roger Jenkins
captain hook	- Pete Billsmith
william kidd	- Brad Cox
jean laffite	- Gerry Spalding
ahab	- Unknown
	(barred from game for misconduct)

"Well there he blows," Bennett said, his voice riddled with accomplishment.

"Hmm...But Ahab wasn't a pirate."

Bennett fell back into the sofa. "I don't think he cared. That was his name and he would keep it whether it fit or not."

"This is fantastic. I knew you could do it!" Hoosier gave Bennett a kiss on the cheek.

Bennett shook his head. "We've still got a long ways to go, but this is progress." His face went straight. "I think I've discovered who our one-armed mystery man is."

Hoosier's eyes popped-wide. "Who?"

"Paul Kitt, Patrick's best friend."

Hoosier scrunched her face. "So does Patrick know anything?"

"I confronted him about it."

"And...?"

"He denies it. But he's edgy, so I find it kind of hard to believe."

Hoosier smiled knowingly. "I shouldn't tell you this, but he's edgy about something else. He shared with me that he had been dynamiting tuna—"

"Tuna?"

"Don't let him know I told you."

"Tuna?"

Hoosier jumped up from the sofa. "Let's celebrate our find. Any wine?"

"I agree we should celebrate, but not with wine." Bennett picked up a brown box—a Hayes 1200-baud modem. "Dang times have changed."

"Huh? No wine? Then name your poison."

"Coffee, black—it's going to be a long night."

The constantly changing intensity of light on the wall above Hoosier's head settled into a singular pattern. She sat slumped in the recliner, deep in sleep, the remote to the TV having long since slipped from her hand to the floor. The test pattern on the TV created the static pattern of light above her head. Bennett took notice, surprised that in the age of 24 hours a day information, that a network went off the air ever. He glanced over at Hoosier, her content face illuminated by the light off the television screen. He picked up his coffee and realized he'd done so for the third time with an empty cup. The notepad lying on the coffee table in front was saturated with notes and diagrams. He had managed to make his way through most of the floppy disks. Only a couple remained, and he intentionally put them last after taking some time at the outset to triage the disks. The most critical ones he would review first and the least significant he would leave for last, so if he fell asleep before reaching them the harm done would be minimal.

Running his fingers through his hair, he took the opportunity afforded him as a result of the distraction presented by the TV to analyze what he had learned so far. His first goal was to understand how the code operated functionally. He traced the operations within source files, and between source files, as necessary. As he did so, he began the process of assembling the various pieces into a framework, which when completed, would define the operations of his programs. He couldn't help shaking his head when reading comments in the code and the coding style, realizing that some of his latest work contained some of the very same nuances. The code was his; he was now convinced of it.

Within the framework he had already established, he had a sense of the breadth, depth, and complexity of what he was dealing with. He found himself quite proud of his accomplishment, not in discerning the code, but having written it in the first place. The components created from his code liberated the typical PC game from the confines of a single machine, or even a network, to the entire world, or what part of it was on the Internet at that time. Granted, a comparatively tiny world by modern standards, but the game would run equally as well today; which in fact it was, but unfortunately in a sick, distorted way. What the game lacked in visual flash, it compensated for by offering tools that aided the mind in unlocking its imaginative potential.

He stretched as he stood and then headed for the kitchen, hoping to find some hot coffee. As he walked, he thought over his next step. He considered Mocha Dick and the assumption he had made that the process carried critical data. His study of the pirate ship code had reinforced the assumption. The process built from the code could hold a large amount of data, probably more than the killer would ever want to stuff it with. It could feed itself files, text or binary, with the aid of a simple script.

He found the coffeepot empty and began the process of making more. A haloed moon greeted him from the dark skies outside the kitchen window. He rotated his head slowly, stretching his neck muscles. If the assumption was correct that Mocha Dick carried important information, then getting information from him had to be the next issue for Bennett to tackle. He would need to find a way to exploit his own code, but should he find that not possible, or not possible without great risk of detection or perhaps accidentally killing the Mocha Dick process, then from what he determined from his analysis so far, he had but one choice, play the game he could not remember playing, and play it flawlessly.

The game itself, he concluded, was by current standards very simplistic. Each player started with a pirate ship and no treasure. Apparently, another participant, the frigates—which Bennett could not locate source code for—carried treasure through various routes, principally from the east coast to the west coast. The goal then, for each player, was to overtake the frigate with his pirate ship and loot it for all its treasure—a low-resolution graphic of some sort. Once the player accomplished this, the next step was to hide the booty somewhere so that others could not steal it. These were known as treasure chests. Once a player amassed over a certain number of treasures, he was eligible for another pirate ship. If one pirate ship overtook another, the winning ship was able, in addition to having what was on board, to retrieve the keys and maps to the loser's treasures and eventually confiscate them. The ultimate winner was the one with the largest empire.

It was a simple game with simple tools. But one requiring skill and a command of the environment in which the game was played. After looking through the update file, where users submitted their changes to the game, Bennett realized just how much the game had evolved in its intricacies. He himself had added code—a module he called chameleon—allowing a pirate ship to hide itself by mimicking the appearance of standard pieces of the operating system where it dwelled. It immediately made him wonder if the system data he had viewed during his checks at Pequod really was what it appeared to be.

The game lost its geographical awareness without access to networks all across the country. Hence, penetration of those networks was necessary to fulfill the spirit of the game. However, the exploits, the tools of the game wielded to penetrate networks, had long since been patched. Patched as a matter of fact, in quadruplicate by routers and access control lists, operating systems, firewalls, and software patches. So how could the current, and only player, still be accessing networks?—Griggs' network for example. It could be poor security and a dozing Unix administrator. But then how to explain Pequod?—locked up more tightly than most networks. He could see how parts of the game would still work, but it was getting those parts delivered and installed that appeared so implausible with the current security measures in place at most sites. Perhaps the player upgraded the code to institute some of the latest known vulnerabilities. But he had missed so many obvious details in other areas, why be so keen here?

Bennett placed his thinking in abeyance, aware of its increasing degradation, to take a few sips of coffee and observe Hoosier. He noticed her goose pimpled arms wrapped about her chest and grabbed a quilt from the back of the sofa, stretching it from her toes to her shoulders. She opened her eyes just a sliver and then shut them. Bennett stepped out his front door to look up at the moonlit hazy skies. He found himself accompanied only male field crickets that serenaded him and the clickety chorus of frogs. In the moment, most of life paused, and to Bennett, brought back the idea that there were still times when even chaos reposed.

He felt his mind wandering, pulling like a dog on a leash, determined to investigate the curious things it sensed. The references in his files to FidoNet and BitNet caused him to chuckle. They made him recall the good old days, when there was some sense of innocence and community on networks. He wanted to recall other fond memories, but time would not permit the tangent and he yanked his thoughts back to the problem at hand.

A game he had innocently devised, to allow he and his friends to play from the comfort of their homes, had been hijacked and hacked into an electronic sepulcher for the victims of a maniacal mind. The thought of it all simply amazed him. The man, who called himself Ahab, came to mind and caused Bennett to shiver. The name kept cropping up, even from the past, and somehow Bennett felt he was near, and had been for sometime, remaining hidden in the shadows of his life, like a bastard brother secluded in the closet, never having access to the light of day.

The night before, after the police officers had returned to their vehicle and Laykin stood at an angle at the end of Granddad's driveway, his clothes disheveled, blood dripping from his mouth, he had asked Bennett one last question. He had asked Bennett how he could be sure he wasn't the killer. After all, there was the odd family history—Laykin had discovered this when he interviewed Granddad hours before the heart attack, perhaps a contributing factor—and he suffered from blackouts during his seizures, something he must have learned from Tanya—how did he really know what he was doing during those periods of time? Bennett shook his head as he thought over Laykin's words—there was no way. But the code was his and the killer was using it and his boot was found…ridiculous presumption, completely ridiculous. And yet the words came back to him.

"How do you know it really isn't you?"

Chapter XIV

He grabbed the can of Coke from the vending machine, wiped the top with his shirt, took a long gulp. A couple of fingers of bourbon and he could be sitting pretty, even while in the depressing confines of the hospital. A few more gulps and he walked from the break area back to Granddad's room. It struck him that according to Bennett he was working for a serial killer. He felt his I'm-getting-into-deep-do-do factor rise. A pudgy elderly woman startled him with her presence as he turned from the hallway into the room.

"Patrick?"

"Aunt Grace?"

The two hugged and Patrick gave her a peck on the cheek. He was glad to see her, but asked how she knew about Granddad. She smiled and said she had a friend at the hospital, but added how nice it would have been to receive a call from one of her nephews. Patrick nodded, telling her he was sorry and blamed it on the confusion. He filled her in on how Granddad was doing, the man now sleeping soundly.

"Although he didn't sleep well last night," Patrick added.

Aunt Grace looked at Patrick and shook her head. "I don't think he was the only one."

"I got in a few winks. Maybe one or two, but who's counting. You know what I mean."

"You doing alright?"

Patrick pursed his lips, folded his arms, and nodded slowly. "About as good as could be expected, you know, given the circumstances."

"I understand."

The two seated themselves in the savanna brown vinyl chairs placed against the yellow colored cinderblock wall. Patrick asked Aunt Grace how she'd been. She commented about life on the island, her garden, some of her friends, the rough ride on the ferry over, and a few other minor items. As Patrick listened he found himself relaxing. If only his life could be so simple and straightforward: going to the grocery store, working in a garden, reading, playing bridge, all so very boring, yet so very attractive to him at the moment. In the last 24 hours he had intentionally ignored the pleas of his conscience to get his act together, to do the right thing, and doing the right thing was a skill that never lent itself to his experiences in life. But as he listened to Aunt Grace, her voice disarming, the pleas grew in volume and duration, and his ability to mute them, lessened.

"Things aren't so good for me right now," he said, before he had consciously decided that's what he wanted to say.

Aunt Grace raised her penciled in eyebrows. "Oh?"

"I don't know how things got out of hand. I've done some things I probably shouldn't have done, and now, Granddad lies ill…"

"What are these things, Patrick?"

"I don't really want to go into it. But, you know, what am I supposed to do?" He saw himself on the sidelines watching as Bennett went to trial, hoping the jury would find him innocent. And if he were found guilty, continue watching as Bennett sat in his cell like a fish on the dock, gasping for life, doomed to perish. He felt himself on a firecracker, a long fuse before him, burning his way—and Bennett resting on a powder keg, a fuse to it burning as well. He could only extinguish one, and right now he was choosing to put out his. Externally, over the years, he had thickened his skin to contain the perceived weakness that burned within, yet now, when the strength he needed lay inside, he could not tap into it, his mind incapable of reaching through to the soft under belly of his callused emotions.

Aunt Grace placed her hand on top of Patrick's. "A person needs to make these things right, Patrick. You already know what the answer is or you wouldn't be talking to me."

"I do?"

"Of course. It just takes courage to do what you need to do. That's what you're really asking me for, courage. Now you and I have not remained in touch, so I don't know much of what you've done with your life, but I will tell you this, everyone deserves a second chance."

"They may not give me one."

"I don't know who they are, but this I do know: it's not them that need to give you a second chance. Patrick needs to give Patrick a second chance. The only thing preventing you from changing your life is you, Patrick."

"But maybe I don't want to change, maybe I like the way I am."

"Perhaps. But let me ask you this, pretend for a moment you're a child again, would you be proud of who you are today?"

His normal reaction would be to act like he was considering the question, while not really doing so, but strangely, he obediently drifted back to his youth, recalling the dreams, aspirations, goals, he thought himself destined for. But they never happened, he had squandered his life in a funk of mediocrity. Worse yet, people he knew he cared for were being sucked into the vortex that had gripped him most of his life. And he knew it was not of their on volition, he had dragged them to the precipice, where they now fought the powerful current attempting to sweep them in. He knew the same means he used to get them there—he could use to lead them away to safety. As a child he would have been ashamed of himself. As a child, his mind would never have entertained his current reputation as a lofty achievement.

Part of his reputation was unwarranted, but was his own doing. He recalled the day he brought the dead dolphin to Bennett's attention, explaining that he and Paul had stoned it with rocks. Bennett believed him at the time, and perhaps still did, but in truth he and Paul had discovered the dolphin beached in the sand further up the Cape, long dead from a scorching summer day.

He sighed. "You're right. I've really fu—messed up."

"So do you know what you're going to do?"

"Not exactly."

"You know Patrick, sometimes the biggest stumbling block to change—and its something you might need to overcome—is guilt. Look into the core of what drives you to act the way you do. It may manifest itself in the way you treat someone, or some group, or how you view life in general."

Hours later, seated quietly in the chair by his granddad's bed, his aunt stepping out to quickly visit some nearby bed-ridden friends, he thought how uncanny she had been. There was indeed something inside, buried as deep as one could dig a hole over a 14-year period of time. For each day, when the thought exposed itself, he dug the hole a little deeper. Funny, no matter how far down he dug, it always reappeared. He recalled the day so clearly: he and Bennett, teenagers at the time, standing on the beach, discussing Bennett taking the boat out for a morning of fishing, the surf crashing on the beach, Bennett afraid it might be too rough. Bennett never recalled what he'd said to him that morning. The trauma Bennett was to experience later, as he watched his father's white shirt—still attached to his back—vanish to the depths of the Atlantic, entrained the memories to oblivion. Though Bennett could not, Patrick could clearly remember what he'd said. And it was those words that he would learn to regret, everyday, for the rest of his life.

The bell on the crusty black tower, which stood no more than ten feet in the air, clanked slothfully in the timid winds off the bay. It stood well secured into Black rock, the huge, nearly flat rock just off the tip of Wilbur Point. Bennett sat on the hood of his car watching the sunrise break over the causeway connecting Sconticut Neck to West Island. He left Hoosier back at the house, waking her to inform her that he was taking a drive. She asked to come along, but he insisted he needed to take the drive alone. She asked when he was going to sleep, he replied in four days. He had told her he couldn't afford to lose time to something so unproductive as sleep. He was only halfway kidding. He would limit his sleep to as little time as he could get by with and still be able to muster enough thinking power to make progress in his search for the killer.

He had overcome his fear of the place, and done so with such ease he had to conclude the fear, like many others, dwarfed compared to what he was presently up against. Why then he was at Wilbur Point, unproductively staring at the sunrise, he wasn't quite sure. Perhaps, he thought, the news article had drawn him, and he wanted to see for himself what the area around the point now looked like. Or perhaps, he thought, he might have a similar reaction to the one he had in his old room at Granddad's and something might focus for him, what exactly, he didn't know, but anything was better than remembering nothing at all. Why this really mattered, in the midst of the countdown, he wasn't sure about that either. But it all appeared to be coming to a head, a sort of confluence of the wounds that had beleaguered his life. Perhaps the impetus for the current situation had actually been set in motion long ago. He couldn't be sure, but what he could be sure of was that his life was being funneled to a single destiny and could not possibly expand again until he came out on the other side a free man.

He looked back over his shoulder at the white gravel drive that circled around the point. In the center, a few two-story cottages and boats occupying most of the space. At the perimeter, one-story cottages, some with beautiful roses, with backyards abutting the rocks of the point. A man and his small dog were navigating the drive. The man looked over at the vehicle with the two men in it. Despite the circumstances, Bennett felt guilty about having the men there. They were following him everywhere and at taxpayer expense. At the same time, his irritation at their constant presence was growing. And then there was the flip side to having them there: the killer could strike again while he was under surveillance, proving it could not possibly have been him that committed the crime.

The man and the dog left the drive and headed straight for Black rock and Bennett. Bennett turned around to face the sea. Moments later he looked down to see a gray Boston terrier sniffing at his feet.

289

"Pickles, come here! Leave that young man alone!"

Bennett turned to see the owner standing twenty feet behind him, increasing his pace to catch up with the dog.

"He's okay, just curious I guess," Bennett said calmly.

The elderly man reached Bennett and the dog, his breathing strained. "Oh…" he said, catching his breath, "beautiful morning isn't it?"

Bennett nodded, squinting as the burning sun brought sparkling fire to the water.

The man looked back over his shoulder. "Who are the two dicks in the car?"

"Dicks?"

"Yeah, you know what I mean, cops."

Bennett shrugged. "No idea."

"I hope they don't mean trouble, we seem to always get that around here."

"You live on the Point?"

"Sure do, going on five years now. The wife and I just love it here." The man bent over and collected an empty beer can. "Except for the damn teenagers. Why they have to hang out here, I'll never know." He tossed the beer can in a rusty black barrel not more than a few feet away.

"I guess it's a long lived tradition around here."

"Could be, but the damn thing needs to end."

The dog had by now meandered out on Black rock, scurrying over it, every now and then stopping to sniff at something. "If you don't mind me saying so, you look like you've had a rough night."

Bennett had not considered that he'd been up all night and was probably looking bleary eyed and unshaven. "I'm in school, pulled an all-nighter."

It seemed to please the man. "Glad to see there's still some young folks out there who understand the value of an education."

Bennett nodded. "So you've been here five years, you say?"

"Yep, five years two months to be exact."

"Anyone you know been here longer?"

The man pulled a white handkerchief from his pocket and bellowed his nose in it. He folded it back up and returned it to his pocket. "A woman named Buella, lives in the pink one over there," he said pointing to Bennett's left, "she's been in that same cottage for over twenty years now." The man spit on the grass and called his dog away from something that had occupied it.

"How interesting. I'm writing an historical paper on the city of Fairhaven, you know, its culture and all, and I need some personal stories to make the paper come to life. Do you think I could speak to her?"

"She may not be your best choice, she's not all there, if you know what I mean."

"Just the same, I'd like to see what she has to say."

"Suit yourself, I'm sure she'll talk to you. She talks to everything else, stones and trees included."

Bennett frowned. "Has anyone else been here for more than ten years?"

"Nope. The damn kids drive everyone away."

The man joined his dog on the rock, picking up what probably was trash. Bennett looked over at the pink cottage and debated. His watch said seven. He decided that was too early to disturb the woman. If he decided later he really wanted to talk to her he could always—

From behind the pink cottage a person, hidden mostly by the powerful sunlight, appeared in the backyard. Bennett covered his eyes and could see someone working in what appeared to be a garden. He lifted himself from the trunk of the car and started walking in the direction of the cottage. When he arrived at the edge of the yard, he spotted an elderly woman wearing a smock standing in a garden, picking small tomatoes from vines towering on cages up to her neckline. She wore a wide-rimmed pink hat and bulky cloth gloves.

Bennett softly said good morning, hoping not to startle her. It seemed to have no effect on the woman, so he spoke a little louder. Finally he practically yelled.

"Don't have to yell son, I hear you." The woman said, continuing with her picking of the tomatoes. "This is a real pisser, the worms are eating every one of my tomatoes!"

Bennett walked towards her.

"How can I help you son?" she asked, still picking at the vines.

Bennett started into a long convoluted explanation of why he needed to talk to her, including references to a thesis and wanting personal stories to throw in.

"I haven't got any stories that would be of interest to you," she said dryly.

"Surely something has happened in the last twenty years that was of interest."

"I can't recollect a single one." She wiped her brow with the rugged glove on her hand. "Boy, this sun is a pisser!"

Bennett tilted his head from side to side. "Well, I'd heard, you know, through some other folks, that about fifteen or so years ago there was a party here on the point that got a little, you know, out of hand. Do you recall that?"

"That's the kind of story you want to include in your paper?"

"Well...I mean, I've got to include the good with the bad, you know, sort of balance it out."

The woman eyed him for the first time. "Who'd you say you were?"

Bennett reintroduced himself.

"I can tell you what I recall, but I tell you, I still to this day don't like to think about it, what a pisser of a night." She walked to the far end of the garden and looked over at Black rock, apparently watching the dog.

"Whatever you can recall, it'd really be helpful for my paper."

"Billy, get off that rock right now!" she yelled. "Damn kid will break his leg, don't know how many times I told him not to climb on that rock."

Bennett could see nothing but the dog on the rock. The elderly man he had spoken to earlier, yelled something out, but Bennett couldn't make sense of it.

The woman shook her head and walked back to her tomatoes. "Kids these days don't listen. They think they know everything."

"I think it was Halloween night? Correct?"

She looked over at him vacantly, pursed her lips, and a light bulb appeared to turn on. "Yes, that's right. They had that loud rock music blaring and trucks full of beer. A lot of them were smoking that weed. Nearly choked me coming through the screen on my back door. I went out on the front porch to give them peckers a piece of my mind. As much as I yelled, they ignored me just the same. I saw a couple of them boys spray-painting something on the Johnson's place and that's when I went to call the police. I'd had enough of those hooligans."

"Did you see anything else?"

"Like what?"

"Fights, someone injured?"

"Come to think of it, I did see something odd."

"What's that?"

She looked back over her shoulder. "Billy get off that damn rock!"

Bennett could feel his frustration level rising. If he could just keep her attention long enough to get the full story. "So what was so odd?"

"What are you referring to?"

"The party that Halloween night, you said something odd happened."

"I did?" She stood in silence, her gloved hands at her sides. "The party, the party...oh, yes, now I remember. I saw one boy dragging another into a car. At first I thought the boy being dragged had passed out from too much moonshine, and for all I know that might really have been the case, but when I saw the blood on his shirt I thought I might have been wrong with my first impression. Once he had the boy in the car, the other boy starting shouting some odd things, then jumped in the car and sped off down the road up the Neck."

"What was it the other boy said?"

"Couldn't understand him, all I know is he sounded like a drunken pirate."

"Pirate?"

"Yeah, you know, that accent they've got."

"What did the two boys look like?"

"Charlie Brown...hell I don't know, what does it matter?"

"It'll just make my story more realistic if I can give some description of the two boys."

"That was a long time ago. Let me see...well, I guess the unconscious boy was thin and wiry, the pirate boy was rather fat."

"Anything more you recall?"

"Nope, best I can do for you."

"One other thing, did you talk to the police about the incident?"

"Police? No. Why should've I? The kid was just drunk, I'm sure of it."

"But they didn't stop by to ask you questions?"

"Guess they might've but I wouldn't have been home. Charles and I leave the Point every year on the first of November to head south to spend the winter in Florida. We left early the next morning, even though neither of us got much sleep that night."

"But you think the one boy might have been seriously hurt?"

"Just saw some blood on his shirt, or I think it was blood, it was very dark where the two were standing."

Bennett thought over what other questions he might ask. He was about to form a question—

"Time to head inside, the humidity is a pisser!" The woman stooped over and picked up her basket containing the tomatoes she had picked and started walking in the direction of the house. When she arrived at her back door she paused and turned towards Bennett. "Don't ask how I recall this, but it seems to me, the injured boy had the letters T R O N on his T-shirt." With that she disappeared behind the door.

As Bennett stood dumbfounded, he mouthed the letters she had spoken. It made sense. In his closet at home, he had packed away the T-shirt he had worn that evening. Why he kept it, he never knew, but he was sure glad he had. Although years had passed since he last brought it out of the closet and had looked it over, he was fairly certain it was a T-shirt from the movie Tron. He now had some answers. He now knew something about what happened that night, the night before the police had found him the next day. Strangely, somehow he knew it had to be the truth, almost as if he remembered it that way.

He sat at a brick red picnic table located just outside the front door of the marina office. The office was closed for the night; the only light remaining, the subdued cast of a soda vending machine parked up against an outside wall of the office. The machine emitted a low hum, the ambience the only companion of Patrick. He sipped his soda, his back to the office, watching car headlights periodically pass by 100 or so yards in front of him, navigating the causeway. With every oncoming headlight he fought off the urge to get up and jump back in his car, and watch as the marina disappeared in his rearview mirror as quickly as he could make it happen. In the preceding hours, a tempestuous debate played out in his mind as he weighed the merits of showing up against not showing up, trying to determine which side of the scale tipped minutely below the other. His mind wavered as rapidly as tin ducks moved in an old fashioned shooting gallery. He didn't wish to leave Granddad alone, or almost alone, he wasn't sure how much company Aunt Grace could really be. Earlier, he twice traveled as far as a mile from the Barnstable Hospital before turning around and heading back. On his third revolution, he managed to go beyond the mile marker and all the way to the marina—the one thought that remained persistent throughout his volley of thoughts, the one that propelled him to continue and not turn back, was that he was the only one who could fix things. And in order to fix things, he had to face the situation head on.

As he sat at the picnic table, continuing to struggle with keeping himself there, he contemplated how he would tell his friend Ahab that he had changed his mind. There would be no final hunt, no final killing of a whale, no training on how to do so. He was done, finished—unable to stomach the mere thought of it all. Ahab could walk the plank. No amount of money could dissuade him. He felt guilty enough over what he had earned so far. The whole thing was sick and he wondered why it had taken him so long to realize it. What could he have possibly been thinking when he first agreed to undertake the challenge?

And now Bennett informs him the man is the serial killer as well.

The situation was too much—way too much to endure anymore. His eyes wandered to the other side of the paved road, to the Lobster Pound, where, like bales of hay, lobster traps walled ten feet high under the purple tint of light emitted from a lamp towering above. He eyed the wood slatted traps, plainly seeing the netting at each end, funneling in towards the middle of the trap like a basketball net. The thick-minded lobster lumbered into the device through the netting, attracted by the bait floating inside and once inside, his puny cognitive resources proved to be his downfall as the tiny brain encased in the crustacean failed to determine that it is able to exit just as it entered.

He had done the same. Of his own volition he crawled into a trap baited by money and the indispensable opportunity to prolong a sadistic binge he and Paul had cuffed nature with since their youth. Any yet he lingered comfortably in the trap for some time, content to squander his days in the enclosure, to gnaw on the bait at his leisure. He questioned the impetus for his sudden urge to depart, was it a conscience, welling up from deep inside, the one he had worked so diligently to suppress, or might it be the sudden jerk on the trap line and the accompanying realization that someone on the surface was in the process of hoisting him from the sea and exposing him to the elements? After all, what AIDS did to sex, time in prison could certainly do to his desire to pursue his current state of self-absorption. But other thoughts, decent ones, had successfully skirted the large perimeter of his ego. And when he obliged these thoughts for consideration, he found their tug a sterner draw than the force behind self-preservation and the desire therein to remain away from prison.

The time had come to leave the trap just as he entered it.

He shifted on the wooden bench of the picnic table. As he did, he felt the barrel of Karl's gun poke his leg. The thought of using it aired with a certain degree of finality, in that one way or another, whether he shot Ahab, or he himself was shot, it would be over, forced to a head—tonight. But he didn't want it to come to that—fervently he swore he would not draw unless drawn upon.

His eyes wandered over to the boats in the marina. Paul came to mind. That night Paul disappeared weighed-in heavily on his thinking as well—definitely a credit in the conscience column: in having one that was. He had relived the night many times. That night they searched in the black, paranoid that any light might draw attention to them, even the weak glow off the red and green lights on the bow and stern proved intolerable, causing Patrick to turn them off. They hunted by sonar and a supportive moon. Clear as that night, he recalled Paul's moonlit body on the stern, heaving his harpoon into a whale, and while reaching for the railing on the bow to steady himself in preparation for the Nantucket Sleigh ride that would shortly ensue, mysteriously finding his wrist knotted in the line scorching across the bow of the *Where II*, and without so much as a gasp, found himself jettisoned into the water where he skied its silken surface for a moment or two when the whale sounded, diving fathoms below with Paul's puppet body in tow, trailing a string of bubbles wrapping his precious last breaths of air. The thought made Patrick shudder—watching Paul's body disappear into the black abyss of the Atlantic—dragged, screaming, to another world, perhaps hell itself. The broken machine was escorted to the bottom of the sea, to linger with other broken machines.

Patrick glanced at his watch. Perhaps Ahab would not show—perhaps a killer such as himself, not bound by the laws of man, found it unnecessary to be bound by the laws of time. His knee began bouncing rapidly with great vigor, so much that the entire picnic table noticeably shook in rhythm with it. He picked up his soda can and sipped air. He crushed it and tossed it in the barrel behind him. Maybe he could accurately conclude that Ahab not showing was a sign—a sign that he would not be meeting him tonight—a sign indicating that perhaps he should leave—

A car pulled into the parking lot, its headlights catching Patrick and pinning his shadow against the wall of the marina. When it stopped, out hobbled an indiscernible figure, a mere silhouette in the dark corner of the lot where the car parked. It approached with silence, in a steady and deliberate pace, headed in Patrick's direction. In the dim reddish glow radiating from the soda vending machine, Patrick saw, when the figure drew to the picnic table, the tortured, stolid face of Ahab.

The strained face, with what appeared as great effort, stretched a smile. "Are we ready for a night of great adventure?—a night fraught with the thrills that accompany the victory in the slaying of the great leviathan? To watch him wallow hopelessly in his watery interment, gasping for air to find room in his blood bursting lungs? How I've longed for this day. Why I didn't accompany you in previous hunts escapes me. I can think of no greater challenge or thrill, save for one, the sweetest taboo in the bouquet of evil inclinations."

As Ahab scoured their surroundings, Patrick spoke up. "And what might that be?"

Ahab pulled his eyes from the boats in the marina and turned them back on Patrick. "That my friend is my secret. One doesn't share such knowledge of such unimaginable treasure with anyone. We are a closed society, those of us who know of it, and our admission standards are by no means meager and not for the faint of heart." A sickly grin, pregnant with condescension, arose on his face. "But! We are not here tonight to discuss treasure; we are here to hunt the mighty whale, the greatest of God's creations. As agreed, tonight we shall take my boat, you need only take me to them, I will perform the delightful task of killing them." He pulled away from the table and began to walk in the direction of the docks. Patrick remained seated. "You hesitate. Why?"

Patrick took a deep breath. "I won't be going with you tonight."

Ahab, in the same deliberate gait he used to first approach Patrick, returned. "Did I hear you correctly? You won't be accompanying me tonight?"

Patrick nodded and slid his hand down over the gun secured in the waistband of his jeans.

"That is by all measures, most unfortunate for you. You see if you're to continue on this course you so stupidly have decided upon, than you will face some very dire consequences I'm afraid."

Patrick swallowed hard. "Such as…" He rubbed the sweat from his gun hand onto his shorts.

"Oh, you're one of those types, the kind that need things spelled out for them. Okay, let me see…how about I completely ruin your credit, have all your property ripped from your greedy little hands? No, no, that won't do it; I've already presented that one previously. Okay, I know…how about I anonymously turn you into the Police for the whale killings? You can't turn them back on me, you have no idea who I am." He read Patrick's face. "Okay, I can see you require further persuasion. Hmm…Okay, how about I kill you? Is that a convincing argument? It is for most I think. But, just incase you think you're special, a martyr of sorts, how about my final option, I kill someone you care about? Candice, perhaps?"

Ashen white fell over Patrick's face.

"Bingo! We have a winner. Now let's just assume I will perform the final option, so without further ado shall we proceed with our fishing expedition?"

Patrick ostensibly reached for the gun, all the while knowing his hand would stop before it gripped the handle. Suddenly every plausible excuse for joining Ahab in the expedition was met with favor. It would be the last time he would have to go, and after all, given his many previous excursions, what was one more?

Something across the street caught his attention. He thought he could make out a woman by the lobster traps, but had to turn his head to capture Ahab. "One last trip and I'm done with it, right?" Patrick said, almost having to yell as Ahab had started towards the docks. He looked back across the street but saw nobody.

Ahab answered without hesitating in his effort to reach the dock. "Yes, that's all I'm asking." He rubbed his hands. "Delicious, delicious…"

The note was short and appeared to have been scribbled in a hurry. It stated that she had decided to go home for a day or two and that there was no emergency, just a need to wrap up some business. After the word *home* it appeared that the words *to Kentucky* had been erased from the paper—the indentation from the sharp pencil used, still lingering. *Kentucky*? The note left by Hoosier made no sense to Bennett. For one, he chose to believe she wouldn't leave now, and two, she never mentioned anything about returning home for any reason. And the reference to *Kentucky*—obviously she had not forgotten she lived in Virginia. It all seemed so strange. He called her home number and got her voice mail message that she would be out of town for a few days. As he mulled through the circumstances, the thought crept in that maybe the stress involved with his situation had overcome her, that she had reached a point whereby in the grand scale of things, her peace of mind outweighed her emotional ties to a relationship that only recently dropped into her life. And that maybe she was not only gone for two days, but for good. It would explain the nervous look of the note and the unwillingness on her part to tell him to his face.

Abandoned. Marooned.

He was alone on his seven-day—now whittled to four—island of freedom. It's beaches eroding, succumbing to inexorable pounding by the surf of fate. He looked over at the shawl he had covered her with the night before. The care he had taken to keep her comfortable. Her slipping away, leaving only a note, cut deep, so deep he felt as if he had been severed in two at the waist. How could he continue?

He had no choice. He looked over at the coffee table, his old PC, the floppies, his notes, spread across its surface. That energy, bursting through his body the night before, was gone; spent on his relentless pursuit to untangle his own creation. He looked at the lipstick on the coffee mug resting by the chair Hoosier had spent the night in. And now his sole source of energy had departed.

He fell, exhausted, to the sofa and stretched out. He lay there for a while, staring at the ceiling. Towards the end, his eyes closing as he felt his body relent itself to the waves of exhaustion that began sweeping through him. Just before succumbing to an unconscious state, he pondered words he had read somewhere before: even a man on death row sleeps the night before his day of execution—and so, four days removed, it was true…

He awoke startled, his mind panicking until it assessed reality, and brought itself back into a conscious state. No Hoosier—the hope that she might return, nestled somewhere in his confused mind, still lingered. He raised himself up and planted his feet on the floor and leaned over the coffee table. He picked up a stack of paper from documents that he had printed-out from the floppies. He vacantly leafed through them and read a line or two on each page. He discovered a page containing what appeared to be an article from some newspaper. Apparently he had typed the article verbatim and placed it in his electronic notes years ago.

"Computer administrators at both MIT and Berkley were shocked to discover yesterday that their systems had been remotely compromised. University officials from both campuses will not elaborate on the

implications to the security of their computer systems, other than to state that the attack was benign and that the perpetrators did little more than to rename the sites. MIT was renamed to "Typee" while the attackers choose the name "Omoo" for Berkeley. Obviously the geniuses who broke into the systems had read up on Melville. A name for a computer system is an address for the benefit of a human who wishes to access it. Computers don't use name addresses they use harder to remember numbers…"

A continuation of the Melville theme: Typee and Omoo were two of the author's first novels to be published. Below the quote, Bennett spotted where he, or somebody else, had typed in the number addresses for the two systems. He found the information of a curious enough nature to drive him beyond his apathy, upstairs to his office where he dialed into Pequod. Proceeding on a hunch, he performed a reverse lookup on the number addresses. The names supplied in the response were not those of MIT and Berkeley. He queried the names database for the Internet and in return received an entire page of information on each name. The page for the first name intrigued him greatly. The owner of the name was Moon Enterprises. He recalled hearing or seeing the business name somewhere but could not pinpoint where or when. On yet another hunch, he performed a search on his PC whereby every file on it was looked through for the name *Moon*. A short list of five files was returned. The first four he recognized as data files for some encyclopedia software installed on the machine. The last file was his inbox for his e-mail. He opened the file and went directly to the match— an e-mail from Griggs.

Bingo.

The number addresses had not changed. MIT and Berkeley no longer had possession of them. It made sense. Back when he and his group were playing the pirate game, Typee and Omoo were two locations on the net where treasures might be buried. The number addresses were either coded into the software and never changed or read from some configuration file that hadn't changed. The new owners of the address numbers were now the new Typee and Omoo. Funny, he hadn't noticed these numbers in the code, but there was so much code that his bleary eyes could have easily overlooked something. The letter from the man identifying himself as the New Bedford killer indicated Mocha Dick was last seen around the island of the Typee. It was consistent with the game.

The intrigue brought temporary relief to Bennett's troubled mind. For the first time in his hellish pursuit, he had a potential trail to the truth. If he followed the yellow brick road, what would he find behind the curtain? The issue before him was how to get behind the curtain. In the day and age of rampant hacking attempts, everybody locked their door in cyberspace.

The days of meandering through someone's site, careful to pickup after one self and to act responsibly when there, were long over. He could try to pick the lock, or call Griggs. But if Griggs had even minimal training, in his position, he would be leery of attempts at social engineering—breaking into a system through people—something Bennett's request for access was surely to be perceived as. It also might alert the team at Moon Enterprises and result in beefed up security. He cracked his knuckles, or tried to, and decided he would go in without the call. Already an outlaw, how much more time could he do for hacking?

Step one was to fire-up his compiler and write a simple program. In order for there to be a game presence at Moon Enterprises, the pirate program he wrote would need to be running on their network. The program he would write would test this theory. And once written he would send an email to the *root* address at Moon Enterprises, with his program as an attachment. If things worked the way he had concluded they should, the pirate program would snatch his program from the email and run it. His program would then send an email to his bogus account, and contained within the email would be system specific information from the Moon Enterprises network. The email would be confirmation that there was a pirate process running on the inside of Moon Enterprises. If he had this, he could then penetrate the network.

Only a few lines of code, he wrote the program in short order. The next step was finding a bounce site. Somewhere to bounce the traffic from his PC off of so that the folks on the other end would think he was coming from the site he was bouncing from. A tour of some of the lesser-known, and very discreet hack sites, where one needed to knock on the door just right to get in, revealed some potential candidates. The first site had cleaned up their act, the second, a government site, had not. Bennett went to work. He began communication with the bounce machine. Once there, he communicated with another server, this one spam-compliant, allowing him to send his email through it to Moon Enterprises. If someone at Moon were to analyze the header of the email and trace its route, they could not follow it back to Bennett.

He waited a few minutes, clicked to his web-mail site where he had established an account based on totally false information. The site was good for his purposes because they kept no logs of access to mail, so there would be no trace of his address anywhere, let alone him ever accessing it. He found his inbox loaded with mail, most of it from users commenting on his serial killer website, but the one marked with the very latest time and date stamp, had a return address of *root@moon.com*.

Bingo!

He found himself smiling. Not only were his assumptions that his foe was using his game and how the game worked, confirmed, but also he now had access.

Darkness now enclosed Bennett. Only the glow from his monitor provided light to his study. He leaned back in his chair and began rocking, time to pause and think through the next step. He believed he had not been discovered and had the luxury of a little time to plan his strategy for once he was inside. He thought through the code used to build the pirate ships, and, he was almost certain, used to build Mocha Dick, and tried to recall if the code monitored the operating system for the presence of suspicious activity. The last thing he needed was to burst onto the scene and have Mocha Dick flee, or kill itself, or kill him. The latter was not his greatest concern as he could always come back, but a computer program never grows weary of repetition and would not tire of killing him a thousand times over. But over time, by means of nothing more than trial and error, he knew he could get around that. But if Mocha Dick were to flee, finding him again could be impossible especially in the short amount of time he had. And if Mocha Dick were to kill himself, then whatever he carried as cargo would vanish, tossed in the proverbial bit bucket. Of course if the machine where he first gained access was not the machine Mocha Dick happened to be on, then he could probe around the neighborhood and stealthily move onto the machine with Mocha Dick.

He got up from the chair and stretched. Sensing his thinking moving in circles he knew a decision had to be made. He knew he had no choice, but to burst onto the scene, for the first machine. He could not recall his software having sensitivity to its environment other than for a few very specific things; his entry would be none of them. He opened the window in his office and allowed the damp air of the night to filter in. In the background outside the window, traffic lumbered down the streets of New Bedford, in the foreground a siren's howl warbled for a short period, leaving a sonorous trail. He settled back into his chair in preparation for the next round.

A siren raced by, followed by a frantic red light flashing in through the window at the end of the room, all of it passing, ignoring her whimpers scarcely penetrating the cloth bit wrapped about her neck and tugging against the corners of her mouth. Hoosier tried pulling her hands apart, but the rope securing them would not give. She had been making the same attempt for a long period of time and could sense her skin becoming raw underneath the rope. She knew she could not give up; her life depended on a successful escape. But the adrenaline that initially flooded her limbs with

Herculean strength was gone, and now her limbs hung heavily at her sides, spent of energy in their futile attempt to free themselves.

Where precisely she was, she could not be sure. Her kidnapper had blindfolded her in the car. He had removed it only after their arrival in the room. Her hands were freed to write a note and she was passed a cell phone with instructions to change her answering machine message. In the darkness, save for the faint light from the moon outside the single window, she could sense she was seated in the corner of what appeared to be an attic. This was confirmed when her kidnapper grabbed a long spear-like object that leaned against a wall, and with a jolly laugh, opened a door in the floor and descended down a set of stairs that unfolded from it. He had told her he would be back after he had slain her sepulcher. She knew who her captor had to be—at least in general terms: he had to be the strangler. Why he had grabbed her—cunningly springing on her at Bennett's house—out of all the possible victims in the entire city, she did not know. The man had to know Bennett, and perhaps his choice was a direct attack against Bennett. She had the very answers Bennett searched for right in front of her, or soon would. If she could escape, not only would she live, but her discovery would be all the evidence required to exonerate Bennett. It was all a matter of devising a plan to get away. She held her stare at a wall until, for that split second when a passing car provided some light in the room, she could see what might be located there. A few cars later, in the corner opposite her, she had discovered a desk with a towering mirror, and a chair in front of it. She continued the process for the remaining areas—the last one yielding something of interest, another spear-like device. If somehow she could scoot her chair over to it…

On the screen were listed the processes running on the machine—Mocha Dick was not one of them. Bennett breathed a sigh of relief. He had sent a program to the pirate program at Moon and the pirate program started it. Because the Moon firewall permitted email (SMTP) traffic, and because his program had hijacked the internal receiving port from Moon's email software, he could communicate directly to his program by wrapping commands in the body of SMTP messages.

Now inside and on a trusted machine, he could stealthily move about the network looking for Mocha Dick. He first checked to see how many machines were on the network and found a small number, somewhere around twenty. A quarter of those could be ruled out because of the incompatible operating systems that they were running, unless his friend had compiled the code to run on them, something that Bennett sincerely doubted. He tarried from machine to machine, poking his head in only long enough to see who was hanging around. After canvassing the network, Bennett's hopes dived to disappointments—no sign of Mocha Dick. He shook his head; it all seemed to fit. Maybe Mocha Dick had left or been commanded to leave. He ran his fingers through his hair, his mind scurrying for grounds to continue his pursuit.

It then clicked.

He had written the module that allowed a pirate ship, or Mocha Dick in this case, to rename itself as a benign process; that being the case, many of the processes listed on the machines could be imposters. He knew of only one way to find out: capture an image, the bits and bytes, of each process and search it for key words he knew existed in his code and therefore would exist in the process. But there were hundreds of processes on each machine—it could take him hours...he needed a script, one that would automate his task and do it in hours rather than days.

He's going to kill me—maybe I should kill him first, Patrick thought, sitting in the stern of Ahab's boat, eyeing the man carefully as he steered the boat towards a remote beach on tiny Brandt's island in Buzzard's Bay. In tow, a dead cow, the one an hour earlier, Ahab had rejoiced in killing. The killing seemed different from the ones he and Paul had participated in. An aura of evil set upon this one like an early morning New England fog bank. Patrick felt as though his skin was crawling with the talons of pernicious spirits. That maybe the vessel, of which he was an occupant, bobbed in hell itself.

At one point they had seen the *Arctic Sunrise* off in the distance. Ahab scowled and took them in the opposite direction. Patrick could not believe he had put himself in a position of potentially getting caught with a serial killer. He looked back at the object in tow, glistening in the dim light of the moon. He looked down at his pants and the tiny piece of chrome from the gun showing through. He had been in the situation before, but could not bring himself to shoot Paul—the idea of even giving it a thought, now brought him extreme guilt—but maybe he could shoot this creep.

A voice yelled out from the bow. "It's perfect—we've got us a floater and high tide is less than an hour away. If we fluke chain him as close to the shore as we can pull him, then when low tide comes, he'll be completely out of the water. I'll return then—"

"Why?" Patrick found himself questioning.

"That my friend, as I have stated over and over again—and quite frankly I'm growing tired of doing so—is none of your business."

The motor was shut off and the two proceeded to row as close to shore as possible. When the boat slid onto the sandy beach with a swish sound, Patrick jumped into the water, removed the line tethered to the whale from the boat and started up on the beach. As he had done in times past, he secured the line around a tree nearby to the shoreline. The two men proceeded to pull the corpse up towards the beach as much as possible. Patrick then took up the slack and wrapped it around the tree. The two returned to the boat.

"She'll sit still until I return," Ahab said, contently.

"You're going to let me go, aren't you?" Patrick asked.

Ahab jerked his head back and smiled. "Of course. Why wouldn't I?"

"I could talk to the cops."

"Stubb, do you think I took my decision to enlist your skills as my whale killer, lightly? If so, you couldn't be further from the truth. I spent months looking for a good candidate. I understand people and can read them quite well. You won't go to the cops, you can't—it's not in you. You could no more go to the cops than I could turn myself in. We're cut from the same cloth, you and I."

Patrick wanted to refute the man, but held his tongue—playing along seemed like a good safe strategy, if there was such a thing with him.

Ahab continued. "I can see in your face you don't believe me. But the fact of the matter is you enjoyed killing these whales or you wouldn't have done it."

"Don't forget you paid me well," Patrick said before he could stop himself.

"Money is not that important to you and if you're telling yourself that's why you did it, then you're fooling yourself. Admit it, you and I are very similar."

Patrick wanted to tell the man his killing stopped with whales, but this time held his tongue. "Maybe enjoy the adventure of it—"

"You enjoyed the killing, stop trying to distance yourself from me. You're a depraved man, Stubb, and the sooner you accept it, the sooner you'll be happy with your life. You need to learn to act on your depravity— live it out. You're a closet killer, Stubb—one who has never acted on his impulses."

Patrick felt affronted by the remarks, but part of him wondered if the man knew something he didn't about himself. After all, it was not impossible that ilk might have remained hidden from him.

"Yes indeed, Stubb, you'll no sooner go to the cops than a banker go to the public with interest free loans."

Ahab directed the boat in a direction other than the dock where they originally departed.

Patrick reached for his gun. "You're not going back to the marina?"

"You can never be too safe, Stubb. We'll return to a mooring I have in another location."

Patrick retracted his hand.

He read the script one last time before sending it. Under normal circumstances a bug would not pose a problem—he could simply fix it and rerun the script. In this situation, however, a bug could set off alarms, followed by his swift demise in the network. Bennett assured himself the script would run without the side effect of announcing himself to the world. Ready to send his script, he began his communication with the Moon firewall, a proxy for the internal mail server, now his program, to indicate that an email was coming its way. After issuing a hello command that falsely introduced himself to the firewall, the following return message came from the firewall: Mail from this domain is blocked.

"Whoa!" Bennett yelled out. From the time he sent the earlier email to now, someone at Moon had configured the firewall to block mail sent from his domain—the part after the '@' symbol in his mail address. His earlier assumptions about Moon had been proven wrong, the administrators were not a bunch of hapless bimbos; they were actually on top of things. But had they really discovered the extent of what he was doing? Or had someone at Moon, watching the logs and spotting an email initiated from his domain, and aware of the reputation of the domain as being an identity withholder, have an *oops* thought, and plug the oversight, right then and there? He couldn't be certain of anything at this point. Either way, his ideas of penetrating the network and operating from the inside were over, not because he couldn't find another email address, there were a plethora to choose from, but because someone was on the ball there and the risk of detection had just shot up tenfold. He would have to conceive an alternative plan.

His mind began a search for his next gambit. The thought occurred to him that if he could not come to Mocha Dick, there might be a way to bring Mocha Dick to him. The seagull processes held the key. If he understood the way he coded their inner-workings, their entire purpose in the game, as scouts, was to assess whether a new piece of geography on the Internet was safe for occupation, and if so, to beckon the pirate ship to port there. The seagulls would use the same mechanism for communication with Mocha Dick as he was attempting earlier, but the difference would be, he himself would not be trying to operate from the inside. The seagulls would use a *pequod.com* domain and he felt confident the Moon firewall would not be blocking the domain. The thought occurred to him that with this tactic, should someone at Moon uncover his game of chicanery, the Pequod name would be in the thick of it. He sighed and shook off his concern, the stakes were too high at this stage in the game; everything he had: property, job, relationships—all were on the table. If he could artificially spawn up a great number of seagull processes at Pequod, perhaps they might gain Mocha Dick's attention and bring him to Pequod. In the code he found one of the indicators that a site was safe, and thus draw the attention of a Mocha Dick, was the number of pigeon, or in this case seagull processes, that could be spawned without immediate detection. Any administrator worth their salt would detect them, so if they went undetected, then the opposite conclusion could be reached.

He thought the idea through and the more he pondered it, the more reasonable it sounded to him. Only one thing prevented him from doing so—*Wolf Spider*. The minute the seagulls started, *Wolf Spider* would trap them and sequester them from the network. He could run the seagulls on his machine, but Mocha Dick, upon arrival, would fall to the same fate with *Wolf Spider*, which would not be such a bad thing if Bennett were physically at Pequod, but he wasn't and didn't particularly want to create a stir by being there. No, the only solution was to temporarily disable *Wolf Spider*, do his deed, and then enable *Wolf Spider* after.

He had little trouble killing the *Wolf Spider* program, which normally might have perturbed him, but under the current circumstances it thrilled him. He decided it best to run the seagull processes on his machine. He started one, communicated with its control port, and told it to spawn ten more.

Rivulets of sweat tumbled down Hoosier's face. The salty drops, which burned her eyes, coupled with the darkness, made it all but impossible for her to see in the attic. She paused her efforts to scoot the chair over by the wall, to catch her breath and determine her bearings. As she labored in her breathing, she waited for passing traffic to illuminate the raven black occupying the room. The door on the floor, a trap door in her path, could be near, and her landing squarely on it and crashing through it to the floor below would not be pretty, nor would it buy her much other than broken limbs. She thought of Bennett and wondered what he thought about the note. Did he buy it? She hoped not. But even if he had found the note suspicious he had no clue where to begin looking for her. Or, maybe he believed it; maybe he believed that she abandoned him. That hurt. She felt sorry for him, now on his own to find the answers he needed to free himself. She pondered his state of mind, how fragile his resolve to continue might be at the moment. If only she could encourage him. But she knew the best thing she could do for him was to escape and bring the cops.

A car passed by and in the split second of light it cast in the attic, she could see the device clearly against the wall, identifying it as a harpoon. If she scooted to her left, she could avoid the door, and if she were then to twist around somehow, she could back into the wall, enabling her hands, tied behind her, to grip the harpoon and slide it down until the sharpened edge on the point was close enough for her to rub the rope back and forth across it, and hopefully sever the rope. It sounded like a plan. She began the tedious motions required to move the chair across the floor.

Patrick stood at the marina and watched as the man called Ahab drove his car from the lot. Pleased that the last trip had not generated violence, something he was sure it would do, he was still surprised the man would let him go. Would it really be the last trip? He had no long-term real life experience in this realm, but everything he had read or heard on the news or seen on TV relating to one individual blackmailing or extorting another, was that the blackmailer never went away, regardless of earlier indications that they might. Why should they? They were obviously reaping benefit from the forced relationship, so why just up and drop it one day? It was not a question of if Ahab would call again, but when he would call again.

A remote chance, but if this were really the last trip, he could walk away, maybe move to some other state, start over, put this all behind him.

Bennett came to mind, evaporating his last transitory thought.

He could not leave Bennett behind to take the fall. He would tell him everything—a decision he made on the boat. He was ready to pay his dues, his debt to society, to get it over with. And when he was out, well maybe do something with his life.

He glanced over at the pay phone standing near the road and walked over to it. He called information and got Bennett's phone number. He took a deep breath and made the call.

"Meet me at the town beach," he said when Bennett picked up.

"The town beach?" Bennett replied blankly. "Can't you just tell me over the phone, I'm very busy."

"No, I'd like to control how this goes down, I don't want them hearing what I have to say."

"They'll follow me."

"That's fine, they won't be able to hear us with the surf in the background."

"What's this all about?"

"Meet me and you'll find out. Name your time."

Bennett sighed heavily. "I don't know, an hour?"

307

Douglas Haydon

"See you in an hour."

Bennett put the phone down wondering what Patrick had to say. Maybe Patrick had played a roll in this mess and was ready to confess. Bennett felt a sick sensation in his stomach. Maybe Patrick was the killer. Something on the screen of his monitor caught his eye. He had started a script that periodically checked his machine for new processes. Oddly enough, the script had found some new processes and they were seagull processes, but ones he had not started. Where did they come from? *What the...*

The harpoon fell to the floor with a loud thud and for a moment Hoosier froze, listening intently, wondering if someone might be in the house and have heard it. She almost laughed when she realized that her efforts of scooting across the floor probably created as much, if not more noise than the harpoon falling to the floor. The levity dissipated with the realization that the sharp edge of the harpoon was out of reach. She would not be able to free her hands. After hitting the floor, the harpoon had rolled down the low grade of the attic floor until it banged against her chair. She could not see it, but felt it strike the chair leg. It occurred to her that although she might not be able to free her hands, perhaps she might free her feet. She moved her right foot to find out how much play the rope would allow. She was able to move it two or three inches—more than she expected. She could do close to the same with her left foot, enough so that her toes made contact with the floor.

The harpoon rested against the outside leg of the chair and she needed it inside. At first she thought of tilting the chair and allowing the harpoon to roll under the right legs of the chair. This, however, if not performed with incredible balance, might cause her to fall over because she had no means to stop the momentum that caused the chair to initially lean. A better plan was to jump the chair in the air just enough to allow the harpoon to slide under. She braced the chair against her back and with every ounce of her remaining energy, drove her body upward. The jump was short lived, but caused the chair to leave the floor just enough for the harpoon to roll beneath it. It did not go perfectly, a front leg of the chair caught the edge of the weapon as it came down and jarred Hoosier significantly, and for a second her ability to remain vertical was in doubt, but through deft acrobatic maneuvers she held herself up—a dividend from three years of ballet in grammar school.

She was able to press the toes of her left foot against the harpoon and hold it in place. The question facing her was where the harpoon point and blade were relative to where she held it. She would need another burst of light. She was aware it was now late into the night and cars passing by were becoming less and less frequent. But she had no choice other than to wait. As she waited, she knew when a car did pass, she would have that single opportunity to gauge all the following moves she would need to make to move the harpoon with her toes to align its blade so that her other foot could reach it. The sweat on her forehead was near dry when she heard the engine from an approaching car. In the light she could see the blade on the harpoon approximately six inches out and upside down. She would need to slide the harpoon back six inches and rotate it 180 degrees. The labor took her longer than she expected, ten minutes or so, but once she had pulled off the moves, she began sliding her right foot back and forth in parallel with the harpoon. She could sense tension in the rope wrapped at her ankle.

The repetitious act tired her, but she would not submit to her body's cries for desisting. Every so often she paused and tried pulling her foot away from the base of the chair. She had no means of judging how frayed the rope might be, or if she were making any progress at all. Her only means of gauging was to test it and hope against all hope that it would give and she'd have one foot free.

Counting repetitions in her head, at the 451 mark, the rope gave. Her foot sprung free. She wanted to take a moment to celebrate, but knew she could not spare the effort—the other foot was still tied to the chair.

Another car approached, perhaps her luck was changing—she would now maneuver her left foot to liberate it.

A car door shut.

He was back.

She struggled for a plan. Maybe a swift kick as he entered the attic—unexpecting. No, it would do her no good. Or she could go back to the corner and hide the fact she'd freed one leg and when she could, work on freeing the other leg. Yes, she might be able to surprise him. Using her free leg she shuffled herself back into the corner.

He did not waste time in making his way to the attic.

As she watched the door to the attic open—a stream of light flickering across the floor—Hoosier had the foresight to thump her foot against the floor in an attempt to make the man think he had jarred the harpoon to the floor as he made his way into the attic. It apparently worked because once he was all the way in, he noted the harpoon, shrugged his shoulders, and placed it back against the wall. Hoosier's heart raced during the few seconds he held the harpoon, her face wincing with each motion he made.

He walked with a determined gait to the desk at the opposite corner of the attic, carefully placing his lit candle on the desktop. He pulled his face from his head; the latex stuck in parts but he gently pried it loose until the springy face bounced as he rested it on the desktop. Hoosier's brow shifted up as she observed what the mask left behind: a rotund section of pasty skin repudiating the warm glow from the candle, and just happened to contain two eyes, a nose, and a mouth. He lifted a wig from his scalp. She could tell he was staring back at her as he looked over his shoulder in the mirror. The crack between his lips arced up. She watched as remnants from his makeover, still clinging like parasites to his head, dangled and dripped. She grimaced as he pulled something from his cheek, balled it, flicked it to the floor, moved to the next string of dangling goop, and stopped. He sighed. "No time for this."

Hoosier started working her left leg, trying to wiggle free by pulling the severed rope through the knot, watching the turtle back of the man as he began a slow process of applying makeup to his face. He paused from his work, and looking over his shoulder said, "I have an exquisite specimen for your burial. You, the ambergris, will rest peacefully there." He returned to his face.

She pulled and felt the rope give slightly, a creaking noise sounded, but when she looked up, she found him preoccupied with applying products to his face. She contemplated how she might get away from the psychotic man. With only one leg to use in her defense, the attempt appeared daunting at best. But the alternative, death at the hands of the man, perhaps finding herself impaled on a rusty harpoon, made the attempt a welcomed one. The man let loose with an eerie guttural laugh. Hoosier looked over as he drew his fingertip down his face, starting above his eye and stopping short of his upper lip. A great smile spread across he face as he rotated his head and studied his features in the mirror. She recalled the man causing Patrick problems was described as having a great scar—

"Let's check in on the great whale tonight and see how he is doing, shall we?" he said, turning to his laptop and clicking it on. As the machine booted up, he returned to the mirror and applied dainty touches of makeup to his face. He lifted a skin cap from the drawer and snapped it to his head. He pulled a gray peppered wig from a sytrofoam bust and fit it snug on the skin cap. Hoosier, observing this from the corner, realized he was making himself appear exactly as he looked when he first came in. He paused with his activity to move his attention to his laptop. He typed a few things and stopped.

"You have not eaten tonight my friend," he said, talking to the screen. "I have fresh food for you this evening, but you did not nibble on last night's leftovers. Are you well?"

Hoosier had not the slightest idea whom the person was he was addressing with his questions.

"Let me check on you..." He typed for a few seconds and stopped. "Where are you Mocha Dick?"

Mocha Dick? Bennett had mentioned the name to Hoosier. The man had to be the killer. Hoosier jumped when the man pounded his fist on the desktop. "Where the hell are you?" He typed furiously, stopping suddenly. "You've left the Typee, why did you leave?" He typed some more. "You were called away?" The keys rattled beneath his fingers. "Yes, you are on your way...on your way to...Pequod!" The man sprung from his chair and began pacing back and forth across the floor. "Damn you!" He formed a fist and slammed it into the palm of his hand. "Damn you!" He stopped in his tracks. "I'm sorry Mocha Dick, you are the greatest of whales, but alas, I must destroy you. I cannot allow my friend at Pequod to disembowel you. If you are to die, you shall die in dignity by my hand, not the plebeian hand of a bumbling fool. I'm afraid you shall never make it to Pequod." He typed out a line and paused, as if reading it over a couple of times and then struck a single key and sat back in his chair. His hands fell to the tops of his legs and he rubbed across them. He returned to the mirror.

His eyes turned on Hoosier. "You are bound for glory, unfortunately not Mocha Dick, but I shall make another like him, one befitting your feminine charm. It's just a matter of attaching my beard and I shall become him, the great unshakable Ahab. Then the ceremony can begin." With that said, he pulled the hairy skin from the drawer and began its application to his face.

The message flashed on his screen for a second time, causing Bennett to shake his head: *Spanish navy in pursuit!* And then it occurred to him—it was part of his game. The seagulls were informing him that a Spanish Navy ship was converging on Mocha Dick, and Mocha Dick being a converted pirate ship, would be destroyed if overtaken. Bennett rapidly went through all that he had learned about his game. What were the evasive actions he could take, surely the game provided some means of escape. But what was the escape?

Spanish navy close!

Something occurred to him. He wasted no time in starting his own pirate ship process. In fact he started five of them. He sent them on their way to the Typee.

As his mind frantically pondered maneuvers, it paused, the irony of playing a kid's game to save his life asserted itself.

Pirate ship destroyed!

311

He prayed it was one of his decoys. He checked, and it was a decoy, the strategy might work.

Pirate ship destroyed!

Another decoy obliterated. He started five more. He couldn't believe it, but he found himself pleading with the Mocha Disk process to hurry. He started five more. He thought the brute force kamikaze approach inefficient—surely there had to be a better way. In between starting pirate ships he would think on it.

But he didn't have to. His script discovered a strange process. It didn't bear the name of Mocha Dick, but it couldn't be anything else. He quickly snapped a copy of it and wrote the contents to a file. It disappeared from his screen. He typed madly to check on the status of the file.

It was there.

He had captured Mocha Dick.

His hands shook as he applied a harpoon to the file. A series of enumerated files named *payload* (number) were created in the same directory. He determined that all the files were image files and began to sequentially view them. One by one the image of young girls, dead, a whale tattoo on their forehead, appeared on his screen. He recognized Kathy Rueger and the girl from his ambergris file. The next image appeared to be a map. It contained a subset of New Bedford streets. A dashed line meandered its way across the page. Bennett could not determine the significance, but printed it out anyway realizing that it had to be of some importance or the killer never would have fed it to Mocha Dick. He stuffed the printout in his pocket. Three more *payload* image files remained.

"My Harpooner," the next image had in white text at the bottom. Patrick stood with folded arms and a crooked smile. "You bastard," Bennett said under his breath.

"The whore," in white text at the bottom of the next, and above, what appeared to be an image of an old Polaroid, in color, of a woman with pancake stringy blonde hair, haloed by a festoon of daisies. She held out her arm, making a peace symbol with her hand, a joint burning in the V. Her eyes were slits and her smile brazenly inebriated.

The last image caused his heart to stop.

Smiling back at him, from a photo taken long ago, was the angelic face of Hoosier.

Without touching a thing, including most of the floor, Bennett flew down the stairs and to his front door and while pulling the handle, heard voices over his perturbed breathing. He put his eye up to the spy hole and in the dark clearly observed Laykin's features and those of his sidekick. The doorbell rang. It hadn't been a week yet!

"Open up Bennett, we've got a warrant for your arrest. Your mine, baby!" Laykin's stern voice hurled through his door. Bennett looked around his home, trying his best to make a good decision with lightening speed.

"We're prepared to kick in your door..." Laykin yelled, in yet a sterner voice.

Chapter XV

"What's going on, my boy?" Alfred asked, Bennett standing in his doorway, heavy breathing consuming so many of his body's resources that speech was all but impossible. "You in trouble again?"

Bennett stepped inside the door and shut it behind himself. His sprint had taken him out his back door, over the neighbor's fence, three blocks down the street behind his home, and two blocks north to Alfred's home. Why Laykin had placed nobody behind the house puzzled him. But he didn't have the time to contemplate why he had been so lucky; there was plenty of time for that later. His legs shook noticeably, to such a degree in fact that his upper body shook with them. He leaned over his knees, placing his hands on them.

"I have...never run...so hard," Bennett forced out between breaths.

"How can I help you, son?" Alfred stood motionless, eyeing Bennett questioningly, not sure how to react.

Bennett stood up. "Can I borrow your car?"

Alfred raised his left brow. "The law after you again?"

Bennett shook his head. "No, I just desperately need to talk to my cousin. My car's not working." Why he lied, Bennett wasn't sure. Maybe the time he felt it would take to explain was time he couldn't afford to spare.

"I'll have to check with the wife—"

"Alfred you have no—" Bennett stopped himself midstream. Again, confronting Alfred over the issue might only prolong his wait. Bennett forged a smile. "I'm sure your wife will be fine with it—it's for a good cause."

Alfred clapped his hands once and winked. "You know what? You're right."

Moments later, Bennett found himself tearing around a corner at Highway 6 onto Sconticut Neck road in route to the Town Beach. The butterscotch Dodge Dart rattled, the steering wheel vibrating in his hands like a jackhammer, and pulling to his right into oncoming traffic. He prayed no cops had taken it upon themselves to patrol the road that night, that no animals would dart in front of him, that the crate he was driving would not blow a tire. The wind howled by his window, while moths and other nocturnal creatures smattered his windshield. It occurred to him that Patrick had left Granddad, but he couldn't worry about that at the moment. He turned down the causeway and flew over the short bridge. Dirt spiraled from the rear of the car, as he turned left, down the first of many unpaved roads. The eyes of a cat illuminated from behind rusty trash barrels placed by the side of the road. He swerved to avoid an ill-placed children's bicycle, on its side, halfway into the road, leaving its front wheel spinning in his wake. Two turns later he approached the chain-link gates that formed the portal for traffic to reach the beach. The car slid for a good ten feet on the sandy pavement used for parking, busting through the surrounding low wooden log fence, sending a timber rolling into the beach dunes.

Bennett sprung from the car and immediately began yelling out Patrick's name as he ran along the parking lot, looking out at the beach as he did so. He spotted Patrick's car parked beneath the tower, hidden from anyone passing in through the gate. He ran in that direction until coming upon the main path, a path winding through the sand dunes, leading to the beach. He dashed into the sand, his legs reacting spastically to the shifting terrain. Clearing the last dune, and reaching the expansive beach, he spotted a figure wading in the surf. "Patrick!"

"Bennett?"

"The killer—where is he? He's got Hoosier." Bennett had reached Patrick and was shaking him.

Patrick's faced glazed with shock. "How did you know?"

"Who is the man?"

Patrick paused.

"Crissakes, Patrick. Hoosier's life weighs in the balance."

"I don't...I mean—"

Bennett forcibly shoved Patrick back, causing him to stumble but remain standing. "Damn it, Patrick! If I have to beat it out of you—"

"That won't be necessary!"

"*Who* is he?"

"I don't know. I swear to it, I don't know. I have met the man, but all I know is that he goes by the name Ahab."

"Where does he live?"

315

"I don't know. He always determined where we met and it was always in different places."

"You killed the whales, didn't you?"

"Yes, Paul and I did."

"And framed me..."

"Frame you?" Patrick lowered his head. "Yes...but—"

"The head injury you gave me, wasn't that enough!"

Patrick's jaw dropped. "Damn Bennett, that wasn't me! You can't blame me for everything that's gone wrong in your life. Sure I was pissed over what you did to Uncle Jack—"

"You mean my father! Don't you think I've created my own hell for myself—I don't need you to contribute!"

"Yes, but the truth of the matter is—"

"What?"

Bennett followed Patrick's eyes to the dark slot near the top of the tower. A glimmer of something flashed in the slot. Patrick pointed his finger at the sand in an exaggerated fashion. Bennett looked at him curiously, paused for a moment, and broke for the sand—face first. *Take cover.*

The shot rang out.

As Bennett lay with one ear buried in the sand, he heard with the other a strange sound, like a shovel slid into soft clay, followed by a gasp.

Four more shots rang out, all piercing the water with distinct plops. No more shots followed.

Bennett spit sand from his lips. "You okay, Patrick?"

No response.

"Patrick?"

No response.

Bennett heard his swallow it was so raw. "You okay, Bennett?"

The sound of tires screeching on gritty pavement flew from the tower. Bennett rose to his hands and knees. "I think he's leaving Patrick. I'm gonna follow him—" He noted the dark pool collecting in the sand next to Patrick's side.

"Go chase the bastard, Bennett," mumbled as the words gasped from his dry lips.

"You've been hit—"

"Hoosier will die. No reason two should die when you can save one."

Bennett glanced back at the parking lot. "I'll never catch him now."

"The day...the day you went out in the boat...the day your father died looking for you...you don't recall this, but I do...clear as I feel life draining from this body...I dared you to go...you didn't want to—you wouldn't have if it weren't for me. I called you a chicken. It's my

316

fault…your dad's death was my fault. My way of dealing with the guilt was to blame you—I couldn't live with myself otherwise. Turns out, I guess I won't have to—" A chuckled coughed from his lungs.

"Bennett!"

Bennett looked over his shoulder to see a figure approaching them from the path.

"Dear God," Patrick groaned, pulling his hand from his side and watching as a black syrup dripped through his fingers.

The figure yelled out again. "It's Agent DeWald. Are you okay?"

"He shot Patrick."

"Who shot Patrick?"

"The killer, he was hidden in the tower. Didn't you see him?"

DeWald had now reached the two. "I saw a car leaving as I entered, but I didn't recognize it. If you hadn't sped so badly, you wouldn't have lost me." He looked down at Patrick, who spit blood from between his teeth.

"But how did you find me?"

"It was Laykin's mistake to assign me to the back door of your house."

"We've got to get Patrick to a hospital."

The two moved over to Patrick.

"He's got Hoosier," Bennett said, carrying Patrick's feet as DeWald carried his head and shoulders.

"Who's got Hoosier?" DeWald grunted.

"The killer."

"How do you know?"

"I just do. I'll explain it later."

The two walked as quickly as possible, moving like a sideways-gaited dog. They reached DeWald's car and placed Patrick in the back seat. DeWald got on the radio to inform the hospital he had a gunshot victim and that he was on his way in. He declined the offer of paramedics, indicating he could get to them faster than they could get to him. DeWald motioned for Bennett to get in the passenger seat.

"I'll follow."

DeWald hesitated. "You will follow, right? My butt is in the slinger as it is."

"Yes," Bennett said, nodding frantically. "Let's go!"

DeWald peeled from the parking lot with Bennett in pursuit. Bennett slapped his hand against the steering wheel out of frustration. He was concerned for Patrick, but in a panic over Hoosier—the two emotions forming a potent flow of adrenaline flooding his muscles with primordial powers. He had to find her.

Douglas Haydon

The map! The thing he printed out. He pulled it from his pocket and turned on the overhead light. Steering with his knee, he looked the map over. It could be. Only one way to find out—Bennett turned a sharp right off of Highway 6 to follow the map while DeWald continued in route to the hospital, cussing wildly, Bennett was sure. He might need help searching the area. He dialed Collin on his radiophone. He immediately got Collin's voice mail. He left a message for Collin to meet him at the location on the map.

Bennett shot to 195, headed west, and turned off right, down Belleville Avenue and eight blocks later, turned left on the street with the name that caused his skin to crawl: Coffin Avenue. The Dart bounced violently as he pulled into the deeply pitted grounds at Riverside Avenue.

The dashed line he followed stopped here on the map with a big red X. He wiped his sweaty brow with a trembling hand and began surveying the three story brick buildings surrounding him, now somewhat illuminated by the morning sun. Windows with few broken panes were left intact while those damaged more seriously were completely boarded up with plywood—skinned in brilliant letters and icons from graffiti. The buildings were all that remained from a flourishing textile industry that brought New Bedford into the industrial age. The brick structures were merely tombstones to the death of the industry. A painted black strip of two-lane girth, bearing in white letters the last name of the once proud owner, a mercantilist, banded the entire brick building.

Bennett combed the grounds for cars or other signs of life, but discovered nothing. He walked up to the nearest brick building and tried to peer into a first story window. Eroded brick crumbled beneath his hand when he placed it on the windowsill, as he looked through the nearly opaque glass in the pane in front of him. He could see little inside, only the outline of shapes, nothing to indicate any recent activity inside. He spotted a door some twenty feet down the side of the building and headed toward it; it was boarded shut. The smell of stale beer and urine crept from the building foundation as he walked through an alley that separated two of the leviathan factories. A couple of rusted kitchen chairs with pallid yellow vinyl cushions, the original pattern, stained beyond recognition, were placed around a rusted trash barrel that breathed remnants of stale smoke from trash once burned inside. Bennett turned from the alley and walked along the backside of the building he had earlier approached from the front. He was relieved to spot Collin's empty pickup and pleased to know his friend had already begun to search.

Despite knowing Collin was looking, he was beginning to feel the oppressive hollowness of futility. The area alone had at least five of the long brick buildings—most abandoned and if he searched them all, finding nothing, what then? Search the entire city building by building? Time was running out—his brute force approach to finding Hoosier was a stupid one. But she could be here—he had to believe that—because it gave him some hope, something to anchor his mind and prevent his thoughts from drifting into the murky waters of his sleep deprived mind. He found a door at the back boarded just as the door at the front, but with one catch: he found he could pry the board away from the doorsill enough to squeeze his body through.

He stepped into a parlor thick in dust and stale air. The pungent dank interior of the extinct factory overwhelmed his senses. He pushed the board back to get some fresh air and to shed some light on the floor directly in front of the door. He called out Collin's name but received no response. He could see footsteps, but could not determine if they were recent. They lead down a hallway and up a staircase. Before following them, he scouted the bulk of the first floor, his eyes slowly adjusting to the dim light resulting from the milky windows. He found nothing to help him in his search. The splinter-bared wooden steps creaked beneath his feet as he judiciously climbed his way to the second story. When he got there it looked much the same as the first. Minutes later, sweat dripping liberally from his face; he reached the top floor, unventilated because no windowpanes were broken and deliriously hot from a sun baking the day before. The top floor consisted mostly of open space; long piles of scraps of cloth, wood, machinery, forming a mountain range relief with the floor, a few rooms huddled at the far end.

Sitting on a low pile of pallets, Bennett rubbed his hands over his face. He debated whether or not to yell out her name. It could expedite her death or bring him to her. He couldn't decide. Thinking clearly as of late was becoming more and more of a difficult task. There were the other buildings. Maybe it was time he stopped, went back to the car and called the cops. As he thought, he listened. The expansive room echoed the tiniest of sounds including the tick of his wristwatch, bringing to the forefront the ever-present sensation of time passing—passing while he sat and she died.

He froze and cocked his head for the faint sound of a beep in the room. He knew his watch did not beep. He stopped his breathing, listening carefully for the sound again.

—Another beep—followed by two more. A ubiquitous sound, one he heard all the time from computers. The fact that the beep continued, at constant intervals, was consistent with the alarm a laptop computer emits when power is waning from its batteries. A laptop in the building was as out of place as solar panels on a stagecoach. Someone had to be here. He had searched the entire floor, walking in silence, skillfully avoiding an encounter with floor debris. Nothing.

He lowered himself on the pallets, staring up at the ceiling. A line of dust fell from in between floorboards. Another line plunked on top of him. He rose from the pallet, noiselessly, concentrating his senses, like a hunting lion crouched for the spring and kill.

There had to be a way up there. He began the arduous task of scanning the ceiling, inch by inch, and eyeing windows for outside steps. In one of the rooms at the far end, he found a cut in the ceiling, square shaped, a handle and unsecured Master lock dangled from one end. He reached and grabbed the handle, pulling it towards him. It pulled freely for a few feet before the rope behind it snagged and as he pulled further, the square section in the ceiling began to separate itself from the ceiling. A wall of dry air greeted his face. As he pulled further, a folded ladder appeared on the topside of the panel. He grabbed the ladder at the base and pulled it to the floor. With a deep sigh and slight hesitation he began his ascent up the ladder.

As his head cleared the ceiling, the dark room above appeared as little more than a small window that floated at a distance in front of him. He paused, waiting for his eyes to adjust to the black world he now found himself entering—the atmosphere in the room, veiled in cunning silence, that stark nothingness attributing to the tingling of one's skin. The beep, this time far more vocal, echoed in the room, accompanied by the short-lived burst of a tiny red light. He jerked his head in its direction. Cautiously he climbed into the room, turned, and walked in the direction of the red flash. The floorboards creaked beneath his feet as he found his way, his hands extended out front, feeling the air for solid objects. His hand reached a smooth surface and felt along it—he guessed it to be a desktop. His hand continued in its travels until it connected with a rectangular object, maybe an inch or so thick. As he felt about it, he soon realized he had a laptop computer—a warm one. It had a power source—probably battery. He felt along the front and found a flat band that when slid to the right, released the lock on the laptop causing the screen section to spring up a small distance. A vein of light appeared in the gap. With his index finger he gradually pushed the screen section up until it was perpendicular with the base unit. A series of random lines, in a multitude of colors, gyrated over the screen. The activity provided a small means of illumination, enough so that Bennett could see the complete desk and a chair in front of it. At first he marveled at the odd setup—a laptop computer in the attic of a dilapidated textile factory.

Then he heard a sniffle.

The sound appeared to come from a dark corner of the room where the light from the laptop could not reach.

"Who's there?" Bennett whispered.

A sound like the creaking of a chair.

"Who are you?" Bennett asked, again whispering, a slight tremor in his voice.

A soft laugh, beginning in silence, progressing too barely audible— but there. "Hello, Jolly Roger." The voice had a heavy accent, distinctly like that of a pirate's.

Bennett could not see the man. He spoke to a dark corner. "You're the pirate."

"The pirate?" A mocking laugh followed. "You idiot, don't you recognize the voice of Ahab?"

"You're the one playing my game, the creator of Mocha Dick, the killer."

"Perhaps."

Bennett's adrenaline surged. "I want Hoosier."

"You're in no position to want anything. I have a loaded gun pointed directly at you."

"So you're going to kill me."

"Yes, just as I killed your cousin."

"Patrick's gonna make it—"

"Dream if you wish."

Bennett's eyes wandered around the room, looking for something he might defend himself with. Lurking by the hole he crawled through to get in the room, he spotted something familiar—the long shadow of a harpoon. His eyes moved to the laptop's keyboard. He quickly went through the key sequence he would need to punch in to change the screen background to a bright solid white.

"You were not much of a player, Jolly Roger. I easily outmaneuvered you—"

"Then why am I here?" Bennett practiced the sequence in his mind—he had to be fast, lightening fast.

"I allowed you here. I knew when you found Mocha Dick you would find me. Why do you think I included a map? I will admit in some ways I did not wish you to find him—he was a god and my receptacle for the ambergris. But in finding him, I had yet another fallback plan, one that would bring you to me."

"Why?"

"Why? The game, I intend to win this time. When I busted you in the head that night at the party, I thought I was killing you, getting you out of the way so that the game could be mine. Unfortunately you did not die, but you lost your memory, so I still managed to reach my goal. You see I took your measly Pirate game from you and have successfully created an entire world. Can you top that? I have a world where men are whole, where prosthetic limbs are removed and real ones grafted in their place, where additional limbs, such as Pip, are made into the whole being they were intended to be. Women are cleansed of their evil ways, when, like the beak of the cuttlefish, they are encased in the redemptive coat of the ambergris. Can your trivial Pirate game approach this! You did not see fit to allow me to play long ago; you thought my strategy too caustic. So now we play with real stakes, in my world, on my terms."

With lightening speed, Bennett's fingers tap—shoed the keyboard with such adroitness that Fred Astaire's jaw would have dropped. The screen flashed a bright white window, illuminating the room. In the resulting light, Bennett observed the bearded man resting in a chair in the corner, a gun in his hand. With millisecond precision, Bennett had a make on the man's exact position. He slammed the laptop shut, dousing the light, and dropped to the floor. Bullets whacked the wooden walls behind him; light streaming through the holes left behind, projecting just above Bennett's body. He rose to his knees and scrambled to the wall where he felt for the harpoon. A bullet whizzed beneath him, between his hands and legs. He fell to the floor just as his hand latched onto the harpoon.

"From hell's heart I shoot thee!" the man roared. Another bullet spanked the wall behind Bennett. "To the last I grapple with thee!"

Bennett rose to his knees, using his best judgement of where he had last seen the man; he cocked his arm up by his ear, gathered his strength, and thrust the weapon forward.

A sucking sound, almost a dainty sip, followed. Another shot rang out, and thereafter the sounds of a heavy metal object striking the floor.

The smell of burned gunpowder saturated the room and an eerie silence, save for the ringing in Bennett's ears, permeated the air. Within seconds, Bennett heard strained breathing coming from the corner. And shortly thereafter a weak, shrilly voice. "For hates sake I spit my last breath at thee…" The spit sounded more like drool from where Bennett lay. He cautiously moved towards the laptop, pausing, waiting for another shot. But no shot rang out. He increased his pace to the laptop, and when there, lifted the screen upright to reveal the white light coming from it.

He could see the man in the corner. He appeared to be resting comfortably in his chair, the harpoon impaled through his chest. Blood seeped the man's shirt, dripping off his shirttail to a small dark pool on the floor below. He managed a crooked smile. Bennett walked up to him.

"Where's Hoosier damn it!" he spat, his voice shaking, and sweat pouring from his face.

A gasp punched through the man's smile, followed by a series of other gasps, until the gasps erupted into laughter. The man sounded like a steam locomotive getting started on the tracks.

"You think this is funny!" Bennett plunged his hands around the man's throat and began choking him.

"You...still...lose..." The man managed to eke from his constricted vocal chords.

Bennett began shaking him violently. So violently the man's hair fell to the floor followed by his beard.

"Stop...it..."

"Where is she!" Bennett screamed maniacally. Bennett noticed the disguise, and noticed more so that removing it seemed to bother the man. He began peeling caked makeup from the face.

"Stop it..." For the first time the mortally wounded man began to struggle.

"Where is she!"

Bennett picked up his pace; furiously tearing off whatever would come loose from the face. A skin cap popped from the man's head, exposing bushy blonde hair. A smaller nose and tighter jowls appeared. Like an archeological dig, what would he discover? The scar flew to the ground, followed by—

Bennett fell back against the wall—his hands shaking violently and his mouth aghast in horror. He stood staring at the face revealed by his hands, while it stared back at him—all confidence drained from it, the sallow truth bared of it. They eyed one another in silence, until the one in the chair dropped his head, apparently losing consciousness.

Bennett shook the man trying to bring him back. "Collin! Collin wake up, damn it!"

Collin's eyes opened so that a sliver of eyeball appeared. "Bennett, I'm going to die, ain't I?" He eyed the black coagulating reservoir at the top of his belly. "Old buddy, how could you have done this?"

"I don't know what happened to you, but you're sick Collin, mentally sick. And you are going to die, but do one last thing for me before you go, tell me where I can find Hoosier."

With his red satin hands, Collin grabbed the top of his shirt, and with a sudden burst of energy, he ripped it open. A milk white tubular piece of flesh rested against his chest. A spurt of crimson blood splashed across the virgin skin. With great pain and effort he raised the appendage from his chest. At its end, five dainty fingers were balled to a fist. In a raspy voice, between breaths, he said, "Let me tell you somethin', my brother Pip, my twin brother—his fist—it rages through my chest—his rage at our dear old mommy...bless her soul...I'm two shakes of a sheep's tail away from her

323

and hell itself. I'm scared, Bennett. I'm gombee in hell fars—" His head dove forward so that his chin rested on his chest. "Pip, don't leave me...you cain't leave me, Pip," he muttered.

"My answer Collin, you owe me an answer!"

The sliver in his eyes closed, but Collin's scratchy voice, crackled through his parched lips. "Brandt's Island...Ahab brought her there...made her the ambergris." Collin slumped in the chair. Bennett felt his pulse and found none. Tears welled in his eyes.

"Collin, you bastard."

Bennett struggled with the helm as Patrick's boat jarred against the whitecaps. He didn't want to make the trip to Brandt's Island. What he might find could traumatize him for life. But he could do nothing else. He had to know the truth. He had to know if Hoosier were there, inside the whale. He hoped that miraculously she would not be there, that if she were, she would be sitting on a rock waiting for him.

A lone seagull rested in the water, eyeing Bennett as he passed by. The early morning sun glistened on the moisture from a thin fog that collected on Bennett's face. A foghorn slothfully filled the air with its tones. He pondered the human blood on his hands and the fact that he had killed a man. Life at the moment had taken on an ethereal posture. He envisioned himself nearing the river Styx, not far in front, the ferryman Charon breaking the chop and paddling across his path, Collin's port body, slumped in the bow like a tiny beluga whale in an Eskimo canoe, his white hand dangling over the side, tickling the sea.

He stopped the boat twenty feet from shore, dropped the anchor, and jumped in, the water splashing about his waist. He shivered as he stood on the beach, looking either way down the shoreline. He started walking. He gazed down the beach as he went. He encountered a dike of large stones and crawled atop.

From there, at a distance, he spotted an elongated gray object planted in the damp sand. A man walked around the object, stopping for closer observation at some points. Further down the beach, a boy was edging toward the man. Bennett mechanically approached what he knew to be a whale. Seagulls feasting on it sang their perturbed song as they soared overhead, sometimes darting at the man by the whale. Unknown to himself, Bennett's pace began to accelerate, still walking, but his arms joining his legs in the motion. His eyes were fixed on the man, who had squatted at an end of the whale. He seemed to try and push his hand against part of the whale.

The man fell back and screamed.

"Oh my gosh!"

The boy had surreptitiously crept close.

"Get out of here, Brett!" The man yelled as he backpedaled, sand spraying beneath his feet. "Dear Lord!" He fell into the surf.

The boy froze.

"Hoosier!" Bennett yelled. He began to run.

"We gotta go…we gotta—" The man grabbed his son and began running in the opposite direction.

"Hoosier!"

Out of breath, Bennett fell to the sand at the perimeter of the whale. He pulled himself to his knees, his chest heaving. Perfume, a thin scent on the bludgeoning waft of dead sea beast, reached for Bennett's nose. It led him to the sand gritted baleen mouth. He could see where the man had pulled some of the baleen aside. His hand shook as he placed it in the same spot.

A groan.

Bennett flayed his arm back.

Another groan. A strained breathe.

"Hoosier…"

"Help…" in a whisper, "me…"

"Hooiser?"

His hand returned to the baleen. He pulled it to the side, and exposed fine hairs of brown.

"Help me…"

The perfume scent was stronger. Bennett combed the beach and spotted a stout chunk of driftwood. He grabbed it and returned to the whale, placing one end at the top of the mouth to the side of the fine brown hairs, while wedging the other end in the bottom of the mouth. He kicked the bottom of the log, driving it back and driving the roof of the whale's mouth up. The profile of a face appeared beneath the hairs. Now with room, the hair tilt back and the face moved forward. It was textured as a prune, white and flaky. The eyes remained shut, but the thin pink lips moved: "help me…"

"Hoosier…" Bennett felt tears in his eyes. "I'll get you out, if I have to tear this damn whale apart with my bare hands." It was all because of him that she was there, engulfed in sticky maw of the whale. It tortured him. But thank God she was alive.

An eternity later, he had her free. A thin red line crisscrossed her neck, having dug into her flesh. When they opened, her bloodshot eyes rolled to Bennett's and a faint smile broke on her lips. "I made it," she whispered.

Bennett nodded, saying nothing, overcome by it all.

325

Straining, she said, "Guess I'm a damn good actress. I got a foot free, but he still got me."

Bennett shhh-ed her.

"We've got to call the police, I know where he is—"

He covered her mouth with his hand. "He's dead."

Relief came to her face. "Ahab is dead."

"Collin is dead."

"He was a monster, wasn't he?"

Bennett heard the voice, but did not turn in its direction, leaving his eyes on the butterscotch colored dirt and small marble like pebbles. He allowed a chuckle, thinking how Collin would have described his own grave: "Looks like the makings of a sundae—peanut butter with marshmallows." The dirt had formed a crust on it. Bennett thought how he was down there in the cool, damp earth, his body a stale empty vessel, the puncture wound to his gut hastily sutured by the embalmer. It was a calm day, warm, and though no flowers conveyed by human hands rested at his tomb, wild ones, buttercups, nestled nearby. The flesh and voice of Collin, that just days before had greeted him at the office, were gone, now beyond the reach of life. Bennett was experiencing both anger and pity. You put yourself there, Collin.

"How is it that we must have monsters?"

Bennett caught Stacey's auburn hair from the corner of his eye. Her thoughtful words from the previous few days had softened his distaste for her earlier actions when she was undercover. He said, "I guess monsters are a fact of life."

Stacey nodded, her eyes on Collin's tombstone. "Hoosier and Patrick doing okay?" Bennett smiled. "Yeah, yeah. Just great. Hoosier gets released from the Hospital today and Patrick will be out in a week. Grandad is already home." He passed his hand over his mouth. "You know there was a part of him that was not a monster." He folded his arms. "I kind of thought of him as a big brother."

Stacey sighed and smiled. "I know what you mean."

She reached her hand into her purse and pulled out a wallet. She spread it opened and flipped through the inner transparent leaves, stopping somewhere in the middle. There she found a small black and white photo of a baby. She removed the photo from the sleeve and held it before her eyes. Bennett looked her way and watched a single tear glisten down her cheek. She flipped the photo over and on the back was written "Your brother." She looked up, away from the photo, and walked to the tombstone, placing the photo at its base. She stepped back and said, "I kind of thought of him as my baby brother, at least for a week. I was adopted. You know I volunteered for the job down here. I knew it would provide an opportunity to look for my brother. I confirmed he was my brother a week ago, but working undercover I couldn't say anything." Tears welled in her eyes. "Such a long search...only to find a monster."

Bennett quoted: "Now small fowls flew screaming over yet yawning gulf, a sullen white surf beat against its steep sides; then all collapsed, and the great shroud of the sea rolled on as it rolled five thousand years ago."

"What does it mean?"

"It's how Melville ended Moby Dick—the ship Pequod, and Ahab, consumed by the sea. What it means to me is God's hand has dealt with Collin, as it has those before him, as it will those that follow." He paused. "It's time to move on, Stacey." He gently gripped Stacey's hand. "Let's go. Join Hoosier and I for a Mac's."

About the Author

Author Douglas Haydon and his wife Sharon live in Northern Virginia. He currently works as a Computer System Architect with a major online service. As a child he spent his summers near the town of New Bedford, Massachusetts, the setting in the book. His experiences when visiting the Whaling Museum in New Bedford are part of the inspiration for the story.

www.ingramcontent.com/pod-product-compliance
Lightning Source LLC
Chambersburg PA
CBHW031820170526
45157CB00001B/126